Floors and Floorcoverings

Richard V. Nunn

Oxmoor House, Inc.
Birmingham

Book Division of The Progressive Farmer Company
P.O. Box 2463, Birmingham, Alabama 35202

ISBN: 0–8487–0451–7
Library of Congress Catalog Card Number: 76–40861

Manufactured in the United States of America

First Printing 1977

Floors and Floorcoverings

Editor: Karen Phillips
Cover Photograph: E. L. Bruce Company
Photography: Andy Hartley
Floorcovering Idea Photographs: Armstrong Cork Co.; E. L. Bruce Co.; Sears, Roebuck and Co.; Louis Joyner; Bob Lancaster; Bill May
Schematic Drawings: John Anderson

Contents

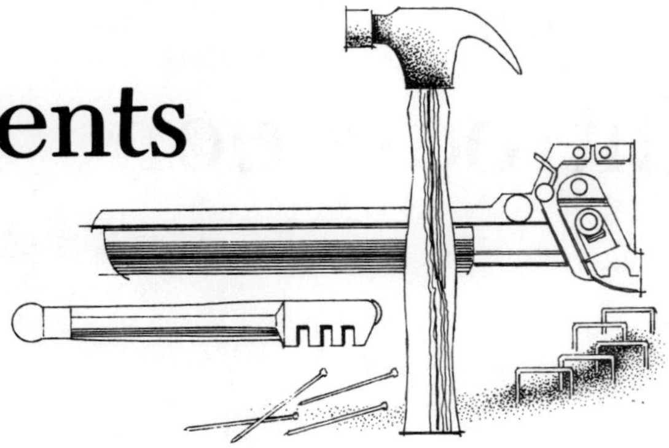

Introduction

Unlike other home improvements (paint and wallpaper, for example), floorcoverings represent a large investment. Carpeting, wood, sheet flooring, ceramics, and resilient tile are expensive. However, the price is not prohibitive for quality products that will last for decades. The key to buying a floorcovering—and installing it yourself—is careful planning and patience during the installation.

Before you shop for any type of floorcovering, we recommend that you take an hour or two to study this book. The information in the following chapters will familiarize you with the types of floorcovering materials available and the problems involved in installing them.

Since the price of some carpeting includes the installation, you may want the store to handle the installation details. On the other hand, you may buy the goods at a carpeting sale and want to do the installation yourself, thereby saving plenty of money. Either way, we think you will find the book helpful since it provides you with basic knowledge about carpeting and how it is installed—professionally or on a do-it-yourself basis.

Resilient tile, ceramic tile, sheet flooring, parquet blocks, and several of the specialty flooring materials are easy for the home handyman to install. In fact, many manufacturers advertise their flooring products by calling attention to the do-it-yourself aspects.

The most difficult floorings to install, in our opinion, are hardwood strip flooring and slate. However, someone with average woodworking skills may use these products; both materials simply take more patience to install than other types of flooring.

Also, before you buy flooring and begin laying it, we want to call your attention to the free literature put out by the flooring manufacturers. There is a wealth of data in this literature; it is available at stores where you purchase floorcoverings and can supplement the basics we show in the book.

Finally, we hope *Floors and Floorcoverings* will help you in making your home a more enjoyable place in which to live. We also hope the book will help you in choosing and installing floorcovering materials.

Facing page:
The prefinished hardwood strip flooring in this formal dining room is the focal point of the room. The handsome floor is further complemented by the oriental rug and the fireplace.

THESE UNITED STATES

Facing page:
A short-shag rug adds softness to this room and draws the outdoor greenery inside.

The orange tones of the terracotta tile floor set the color theme for this garden room. The room opens onto a patio constructed with quarry tiles.

Above:
Unfinished pine boards form the strip flooring, ceiling, and interior walls of this room. The simple furnishings place the emphasis on the natural surroundings of the home.

Left:
Vivid color combines with sculptured carpeting to provide a beautiful textured counterpoint to the sleek woods of contemporary furnishings.

Facing page:
The dark color of the beautiful ceramic tile floor provides the setting for the plants, furnishings, and the airy, grid-type ceiling in this solarium.

The bright gold and white colors of the vinyl sheet flooring in this attic sewing room brighten an otherwise dark room.

The mosaiclike pattern of this tightly woven loop carpeting lends an air of casual elegance; in addition, its wearability makes it an especially good choice for a much-used room.

Overleaf:
The shag carpeting in this room matches the champagne-colored walls and ceiling to provide a neutral background for the vivid colors of the furniture.

Floors in Your Future

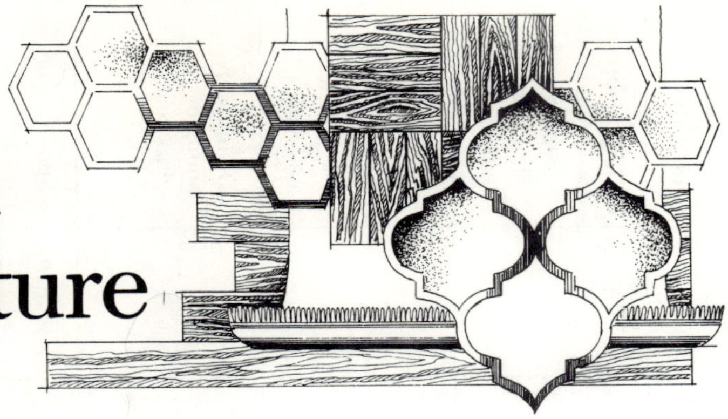

Floors are a peripheral part of your home. Floorcoverings should be subtle and should contribute to the overall mood of the room. Flooring should complement your furnishings, lighting, windows, doors, walls, and ceilings. It becomes a component of a well-dressed room.

If you have any doubt about the variety and availability of flooring products on the market today, visit several flooring dealers in your community. The selection will stun you. In fact, the toughest problem in putting down new flooring in your home may be selecting the product.

We have included a chart to help you make your selection. However, for more specific information, ask your flooring dealer about the wearability, cleanability, insulation, sound conditioning, ease of installation, cost of installation, maintenance, and replacement of the product.

Frankly, we have no recommendation as to the type of flooring that you buy. It will depend on your home. The only guideline we can give you is this: buy quality flooring or it will not hold up over a period of years. If flooring could be easily replaced, we would tell you to buy the less expensive flooring and change it every five years. But labor costs for installation—and your time, if you do it yourself—are at a premium today.

You will be surprised to discover that quality flooring does not cost much more than inexpensive flooring when you consider the replacement factor.

When you buy flooring, consider these points:

- Flooring is a decorative element of your home, and you should buy what you want. There are no hard and fast rules to follow when selecting a product.
- Have several ideas for flooring in mind when you shop. For example, choose flooring that will complement the color of your kitchen or another room in the house.
- The cost may be important to you. But, as we have pointed out, when you purchase quality flooring, you will probably save money over a long period of time.
- Think about the mood you wish to create in the room you are reflooring; both color and texture play an important role. Dark colors make the floor appear less prominent and help shrink the size of the large room. Light colored floors make rooms seem larger. Patterns, light or dark, help accent furnishings.
- If you are buying resilient flooring, consider the ease of installing and maintaining the product. When buying carpeting, consider the type of fiber and its wearability, the ease of maintenance, the construction of the carpeting, and the place where the carpeting will be laid.

Floor and Carpeting Chart

Type	*Relative Cost*	*Ease of Installation*	*Durability*	*Thermal Insulation*	*Traction*	*Range of Styles/Colors Available*	*Ease of Stain Removal**	*Wear Resistance*	*Resilience*	*Maintenance*
Hardwood Strip	$1 to $3 per sq. ft.	Difficult	Life of the house	Fair	Good/slippery when wet and/or waxed	Many wood tones/grains	Excellent	Excellent/shows wear in high traffic areas	Low	Dust mop/vacuum/clean and wax/no water
Parquet and Block	Usually sold by carton; about $1.50 per sq. ft.	Easy	Life of the house	Fair	Good/slippery when wet and/or waxed	Many wood tones/grains	Excellent	Excellent/shows wear in high traffic areas	Low	Dust mop/vacuum/clean and wax/no water
Ceramic Tile	Usually sold by carton; about $7 per 5 sq. ft.	Medium difficulty	Life of the house	Fair	Good/glazed tile slippery when wet	Almost unlimited in colors and patterns	Excellent	Excellent but may discolor somewhat	Low	Damp mop/occasional cleaning with tile cleaner and sealer
Mosaic Tile	$1 to $2 per 12 inch sq. sheet	Medium difficulty	Life of the house	Fair	Good/glazed tile slippery when wet	Somewhat limited in color and pattern range	Excellent	Excellent but may discolor somewhat	Low	Damp mop/occasional cleaning with tile cleaner and sealer
Slate	$40 per 5 sq. ft.	Medium difficulty	Life of the house	Fair	Good/moderately slippery	Limited	Excellent	Excellent/shows some wear in high traffic areas	Low	Damp mop/semigloss or gloss sealer
Vinyl Resilient Tile	50¢ to $1 each	Easy	20 years	Fair	Good/slippery when wet	Almost unlimited in colors and patterns	Excellent	Good/may fade and discolor somewhat	Excellent	Sponge mop/rinse/vinyl dressing in heavy traffic areas
Sheet Vinyl	$5 to $15 per sq. yd.	Medium difficulty	15–20 years	Fair	Good/slippery when wet	Somewhat limited in color and pattern range	Excellent	Good/may fade and discolor somewhat	Excellent	Same as vinyl tile/some types never need wax or vinyl dressing

*See Stain Removal Chart for Specific Instructions

Floor and Carpeting Chart (continued)

Seamless	About $80 per 120 sq. ft.	Easy	15–20 years	Fair	Good/ slippery when wet	About 20 variations	Excellent	Excellent	Good	Damp mop/ no waxing
Indoor/ Outdoor Carpeting	$5 to $8 per sq. yd.	Easy	7 to 10 years	Slight	Good/may be slippery when wet	Limited to colors and patterns	Good/ excellent mildew resistance	Good/tends to fade and discolor outside	Good	Clean with garden hose outside/ vacuum inside
Carpet Tile	50¢ to $1 each	Easy	5 years	Fair	Good	Limited to colors and patterns	Fair	Good/may tend to curl at edges	Good	Vacuum weekly/ overall cleaning: use dry powder or aerosol foams/clean professionally every 3–4 years
Shag Carpet-ing	$6 to $12 per sq. yd.	Medium difficulty	10 years (with pad-ding)	Fair	Excellent	Somewhat limited in color and pattern range	Good	Good/some fading and matting	Good to Excellent	Vacuum weekly/ clean profes-sionally about every 3 years
Sculptured and Loop Carpet-ing	$4 to $12 per sq. yd.	Medium difficulty	10 to 15 years (with pad-ding)	Fair	Good	Wide range of colors and patterns	Good	Good/some fading and tends to wear in high traffic areas	Excellent	Vacuum weekly/ may be spot cleaned/clean professionally every 2–4 years depending on dirt conditions
Plush Carpet-ing	$8 to $16 per sq. yd.	Profes-sional instal-lation recom-mended if you don't have equip-ment	15 to 20 years (with pad-ding)	Good	Good	Wide range of colors and patterns	Good	Good/some fading	Excellent	Vacuum weekly/ clean profes-sionally every 3 years/use dry cleaner or aerosol cleaner on spots

Stain Removal Chart

Stain or problem	*Removal technique*
HARDWOOD STRIP, PARQUET, AND BLOCK FLOORING	
Food	If stain is fresh, wipe with a damp cloth. If old, wipe with a damp cloth. If the spot looks dull or worn, apply wax to it.
Grease, tar, oil, lipstick, crayon, heel marks, paint	Lightly rub the spot with superfine steel wool (not a scouring pad) dampened with mineral spirits. Wax the spot and polish.
White rings and spots	Rub the area with superfine steel wool dampened with wax. If this doesn't remove the spots, try steel wool dampened with mineral spirits. Wax the spot and polish.
Water spots	Rub the area with superfine steel wool dampened with mineral spirits. When it is dry, touch up the spot with matching stain. When the stain is dry, wax and polish the area.
Dark spots	First, try the white ring or water spot treatment above. If this doesn't work, apply household bleach to the spot. Let the bleach stand one hour. With water-dampened cloth, wipe away the bleach. Smooth the area with superfine steel wool, and use a matching stain. Then wax and polish the area and entire floor.
Ink stain	Put household bleach on the stain as described above. Or rub the stain lightly with superfine steel wool, wash it with mineral spirits, and retouch the spot with matching wood stain. Wax and polish.
SLATE	
Coffee, juice, alcohol, lipstick, crayon marks, dried food, mud, dirt, grease, animal stains	Wash the area with a mild detergent mixed in hot water. Rinse the area with clean water. Wipe dry.
Heel marks, paint, shallow scratches	Try the above technique. If this doesn't work, immediately, while the surface is wet, rub the problem spot with powdered pumice or sprinkle the pumice on a water-dampened cloth. After you remove the spot, rinse the area with clean water and wipe dry. Polish.
CERAMIC AND MOSAIC TILE[1]	
Alcohol, coffee, juice, food, mud, dirt, milk, tea, animal stains, ink	First, try detergent mixed with water. If this doesn't work, use scouring powder and a brush.
Tar, grease, lipstick, crayon, heel marks	Use mineral spirits on a soft cloth. When the stain is out, scrub the area with mild detergent mixed in hot water. Rinse the area with clean water. If grout lines are affected, you may have to clean out the old grout with a narrow chisel, or pick and regrout the affected joints. You may also use regular tile cleaner to remove most of these stains.
Candle wax, paint	Remove all wax or paint you can with the edge of a case knife. Then clean the area with scouring powder or mineral spirits on a soft cloth. Scrub the area with mild detergent and hot water; rinse.
VINYL FLOORING	
Alcohol, coffee, fruit juice, foods, animal stains, blood, tea	Use a cloth moistened with mild detergent and water. Wipe floor as quickly as possible after spill. Rinse the area with clean water and dry. If the stain is still noticeable, try lightly rubbing the area with superfine steel wool moistened with mild detergent.

Stain Removal Chart (continued)	
Stain or problem	*Removal technique*
Heel marks, asphalt, tar, candle wax	Rub the spots lightly with superfine steel wool moistened with mild detergent. You may also use special floormark removal pads. Or use a soft cloth moistened with *vinyl* cleaning and rewaxing solution.
Paint, varnish, shoe polish, fingernail polish	If any of these products have dried on the flooring, carefully scrape off as much excess material as possible with the blade of a case knife. Then, go over the spot with superfine steel wool dipped in vinyl floor cleaner or liquid floor wax remover. If the spill is fresh, blot the spot with a cloth. Then wash the area with paint thinner or mild detergent; rinse with clear water. You may need to buff the area with self-polishing or buffable wax made for vinyl flooring.
Heavy dirt and grime	Damp mop. If this doesn't remove the debris, mix ⅛ cup of mild detergent with one gallon of water. Rinse with clear water. You can use an electric scrubber for this. And you may use a wax-cleaner made for vinyl flooring.
CARPETING[2]	
Ink from ballpoint pen	Use spot remover, several times if necessary. You cannot remove indelible ink.
Beer	Blot; then wash with lukewarm water. If any rings remain, use shampoo-type remover.
Blood	If fresh, blot with cold water. If old, wash with mild detergent and warm water. Rinse.
Candle wax and gum	Cover area with cloth filled with ice cubes. Remove all possible wax and gum with your fingers. Then remove remainder of wax or gum with spot remover.
Cocoa and coffee	Use spot remover; then wash area with mild detergent and water. Rinse and blot dry.
Crayon, food, grease, lipstick	Use spot remover.
Juice	Use spot remover; then use shampoo-type remover. Rinse and blot dry.
Grease and oil (lubricating)	If the spot has a crust over it, remove all the crust you can with a case knife or your fingers. Then use spot remover.
Paint	Use spot remover. If the paint is water-based, use warm water mixed with a mild detergent. Or use shampoo-type remover. If the paint is difficult to remove, use both.
Tar	Use several applications of spot remover; then use shampoo-type remover.
Animal stains	Use shampoo-type remover.
Alcohol	Blot. Wash with mild detergent and water. If stain can still be seen, use shampoo-type remover.
Urine	Blot. Mix mild detergent with water and a tablespoon of white vinegar. Scrub and rinse the area. Blot.
Soot from fireplace	Vacuum all loose soot. Use spot remover; then use shampoo-type remover.

[1]Glazed tile is nonporous; therefore, it will not absorb any liquid or dirt. Unglazed tile will absorb some liquids and dirt. The above chart is for nonporous tile.

[2]There are two types of commercial cleaner available: a spot remover and a shampoo-type remover. As a general rule, spot removers are used on non-water-soluble stains. Sometimes, as the chart points out, a combination of both works best. After removing a spot with a commercial cleaner, you may have to have the entire carpet cleaned.

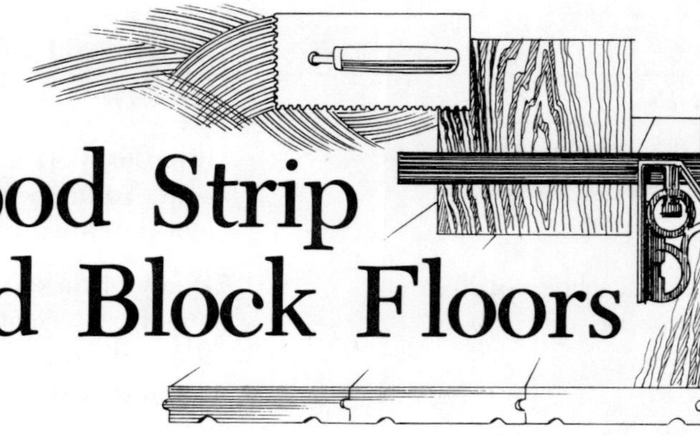

Hardwood Strip and Block Floors

Wooden floors add a certain richness and warmth to any home. Since most strip flooring, parquet blocks, and plain wooden blocks are now prefinished, the task of laying a wooden floor is within an average handyman's skills.

Water and water vapor can severely damage any floor. Before you start installation procedures, you should correct any water problem that you may have. Here is a checklist:

- Water in the basement or under your home can be caused by an inadequate rain-carrying system (gutters and downspouts), cracks in the foundation of the house, or leaky pipes.
- Ill-fitting doors and windows that let in water.
- Excessive atmospheric humidity.
- Green lumber in the subflooring or floor joists.

If your home is built over a crawl space, you can protect the floor from excessive moisture by spreading 4-mil polyethylene film or 55-pound, roll roofing over the ground between the foundation walls. Cover the ground with parallel strips and lap the edges of the strips from 2 to 4 inches. You do not need to seal the seams.

Ventilation under the floors is also necessary to avoid damage to wooden floors. The cross circulation of air in crawl

Hardwood Floor Nailing Chart

Flooring size	*Nail size and type*	*Nail spacing*
¾″ by 3¼″ or ¾″ by 2¼″	7d or 8d steel-cut nails	10″ to 12″ on center
¾″ by 1½″	Screw nails 1½″ long	10″ to 12″ on center
½″ by 2½″ or ½″ by 2″	5d or 6d steel-cut nails	8″ to 10″ on center
½″ by 1½″	Screw nails 1½″ long	8″ to 10″ on center
⅜″ by 2″ or ⅜″ by 1½″	3d or 4d steel-cut nails or screw nails 1″ to 1½″ long	6″ to 8″ on center

spaces (and some basements) is important. If your home already has vents for this, leave them open except during the heating season. If your home is not vented, add vents. The total area of vent openings should be 1½ percent of the home's first floor area.

Preparing a good subfloor

If you are adding a new room to your home and covering the floor with hardwood strips or blocks, the subfloor should be built with 1- by 4- or 1- by 6-inch, No. 1 or No. 2, common-seasoned, thoroughly dry lumber that is square edged. These boards should be installed diagonally across the joists and nailed to every joist with two 10d nails. Leave ¼ inch between the boards for expansion and contraction.

You may also use plywood for the subfloor. The plywood should be at least ½ inch thick, and the sheets should be installed with the grained outer plies at right angles to the joists. Nail the panels at 6-inch intervals along each joist with 7d (or larger) nails.

If you are installing a new wooden floor over an old floor, you still must replace any rotting, splintered, or worn areas in the subfloor. The new floor should be laid at right angles to the old floor.

Wooden floors may also be installed over concrete. This procedure is explained later in the chapter.

How to lay hardwood strip floors

To lay hardwood floors, you will need these tools:

- 13-ounce claw hammer
- Small pry bar
- Hand or portable electric drill and drill bits
- Nail set
- Ripsaw (for sawing flooring with the grain)
- Backsaw and miter box (for sawing flooring against the grain)
- Flexible measuring tape
- Chalk and chalk line
- Flooring nails (see Hardwood Floor Nailing Chart)
- Wood putty or wood plastic

Prepare the floor first. Remove all furniture from the room, and remove the baseboard and any other obstructions from the floor. If the floor is rough, you will have to cover it with an underlayment. Use hardboard or plywood for this, and be sure to mismatch the joints of the underlayment, as shown. In remodeling, board subfloors should be laid diagonally. Leave ¼ inch between the boards, and nail them to the joists with 10d nails. If you use plywood, lay the plywood so that the grained outer plies are at right angles to the joists. Space the nails 6 inches apart along each joist; use special "ringed" underlayment nails.

Unwrap the flooring and leave it for at least 72 hours in the room in which it will be installed. This permits the flooring to adjust to the humidity in the room. We recommend that you spread the strips over the subfloor or underlayment at right angles to the joists. Distribute the long and short strips so that the end joints are not clustered. If you use plank flooring, make sure that the first board in every row is cut so that there is no lineup of pegs along the wall. After you remove the baseboard, space the first strip about ¼ inch from the wall surface, as shown. This allows for expansion and contraction of the wood.

Tack the first two strips of flooring in place. Then snap a chalk line on the subfloor or underlayment to make sure the boards are set properly. Don't worry about the gap at the wall; a baseboard or a molding will cover it. Face-nail the first strip of flooring every 10 to 12 inches along the edge nearest the wall. This way, the nailheads will be covered by the molding. Blind-nail the tongued side of the flooring, angling the nails about 45 degrees where the tongue and the shoulder meet.

Far left:
Use a piece of scrap flooring as a buffer block when you tap the tongues and grooves together. This will protect the flooring from unsightly hammer tracks. The joints of the flooring should be as tight as you can make them. Since the flooring is tough, you may have to drill tiny pilot holes for the nails. The holes will prevent the tongues from splitting as the nails are driven into the subflooring or underlayment.

Left:
Drive the flooring nails at a 45-degree angle through the tongue of the flooring, as shown. Use the Nail Chart to determine the spacing and the size of nails. Countersink the nailheads with a nail set. The nails do not have to be driven into the joists that support the subflooring.

Stagger the flooring joints, as explained earlier, to make a neater looking floor. The layout will give you the opportunity to interchange the strips, stagger the joints, and match different pieces of wood.

To fill out a strip of flooring, run the gap board against the nailed board and scribe a line at the end of the nailed strip. Check the end of the nailed strip and the gap board to make sure these joints are perfectly square, as shown.

If the strip has a tongue on the end that is to fit against the wall, cut it off; you want solid wood up to the junction of the floor and the wall. For doorways and other obstructions, make a cardboard template for proper fitting. Mark the flooring from the template, and cut the flooring to fit.

The final strip is face-nailed to the subfloor or underlayment. You will have to use the small pry bar to fit the last and the next to last boards together tightly. Wedge the bar between the last strip and the wall, and apply pressure. Have a friend face-nail the last strip while you supply the tension from the pry bar.

The molding is applied this way. The baseboard is nailed to the wall at the studs in the wall. The studs are 16 inches on center. The quarter-round molding is nailed either to the baseboard or to the flooring. Nail the quarter-round molding at a slight angle. The molding should be finished to match the wall—not the flooring. We recommend that you give new molding a prime coat of finish before you install it. Then give this molding a final coat, sanding the surface of the molding with a medium-grit abrasive between coats.

Sanding and finishing

Most hardwood strip and block flooring sold today is prefinished so that you do not need to sand and finish the wood. If you buy unfinished flooring, you will have to sand and finish it. Here are the steps:

1. Rent a floor sander and an edger. The cost of buying this equipment is prohibitive.
2. Sand the main part of the floor smooth. Fill any cracks with wood filler. The surface will take at least three different sanding cuts. Use quality abrasive paper. You will need grades 20, 36, and 80 for the large drum sander. For the floor edger, buy grades 20 and 60 abrasive paper. Start with grade 20 and work up. You will also need a sanding block for the edges. Buy both coarse and medium abrasives to use on the block.

   After the main floor has been sanded, use the edger to go around the walls of the room. But before you do any sanding, make sure that all nails have been set, the floor has been vacuumed, and all wax and debris have been removed from the floor.
3. Finish the floor with a penetrating sealer. Before you apply the sealer, vacuum the floor. The finish should be applied to the floor as soon as possible after sanding. This prevents fuzzing or raising of the grain on the unfinished floor surface.
4. When the finish is dry, apply one or more coats of paste wax or solvent-type paste wax to the floor. Buff the wax to a shine with a soft cloth or an electric floor polisher.

Wooden floor maintenance

You should care for your floors with good quality, solvent-based, wooden floor cleaners and waxes.

Never use soap and water on a hardwood floor. Water can damage the finish and the wood, and it can cause the floors to warp. For cleaning, use a high quality solvent-type cleaner—do not use a self-polishing, water-based cleaner. Clean a small area at a time, pouring the cleaner directly onto the floor. Spread the cleaner with a soft cloth; then use a clean, soft cloth to mop up the debris. Now wipe the floor. Let the floor dry, and polish the wood with a clean cloth or an electric polishing unit.

Measuring and preparation for parquet floors

Parquet blocks are wood in block or square tile shapes. To lay this type of floor, you will have to estimate the size of the room. Measure the length and width of the room, and take this measurement to the flooring dealer. He will determine how much flooring and adhesive you will need for the room.

Block flooring must be stored for at least 72 hours in the room in which it will be used before you start to lay the floor. This permits the wood to adjust to the humidity in the room. We recommend that you unpackage the flooring and spread it out over the floor in order to permit air to circulate around the wood.

Parquet flooring may be installed over wood, concrete, or resilient floors. However, the floor must be smooth and firm, or the blocks may break from the uneven subsurface. Do not lay a block floor directly over a hardwood strip floor; install an underlayment of plywood.

Concrete floors. The concrete must be at least 50 days old. If there are high spots on the concrete, these must be leveled with a terrazzo grinder or Carborundum stone. If the concrete is low in spots, fill these low areas with latex concrete patching mix.

Wood or plywood floors. Plywood subfloors must be solidly nailed at the edges and through the center. If there are rough spots on the wood or plywood, sand them smooth. If the subfloor is old, make sure it is nailed properly and sanded smooth. Old surface floors should

HALF PRICE BOOKS®

Half Price Books
1835 Forms Drive
Carrollton, TX 75006
OFS OrderID 39481319

Thank you for your order, ABE TP Orders!

Thank you for shopping with Half Price Books! Please contact service08@hpb.com. if you have any questions, comments or concerns about your order (738040417)

Visit our stores to sell your books, music, movies games for cash.

SKU	ISBN/UPC	Title & Author/Artist	Shelf ID	Qty	OrderSKU
S449358117	9780848704513	Floors and floorcoverings (Family guideboo... Nunn, Richard V	HOM2.5	1	

SHIPPED STANDARD TO:
ABE TP Orders
800 Avondale Ave
Columbus OH 43212
help@textbookrush.com

ORDER# **738040417**
Abebooks

be clean. Remove all varnish, paint, shellac, or wax.

Resilient tile. Some blocks may be laid over asphalt or vinyl-asbestos tile. Check with the flooring dealer. The tile must not be crumbled or loose. The tiles should be firmly bonded to the subflooring. If the tiles are not firmly bonded, remove them from the subfloor. Sand or scrape the subfloor to remove the adhesive. If they are well bonded, sand the surface of the floor to remove the wax.

How to install parquet and block floors

You will need these tools and supplies to lay parquet floors:

- 13-ounce claw hammer
- Crosscut saw
- Coping saw
- Chalk and chalk line
- Notched adhesive spreader
- Flexible tape measure
- Flooring
- Adhesive

This flooring is tongue-and-groove block flooring. It is usually prefinished so that you do not have to finish the wood after the floor has been laid. The floor may be waxed, and some waxes have a cleaner built into them. To maintain the floor, do not use a water emulsion, self-polishing wax.

Lay out the floor, using a chalk line (see drawings). The lines have to be perfectly true; check them with a framing square, as shown. If you make an error, it will be compounded throughout the room. Start laying the units at a 90-degree angle. We recommend that you lay one row of tile down both of these lines from the point where the lines cross. This layout will help you determine your measurements, and it serves as a double-check against incorrect measurements.

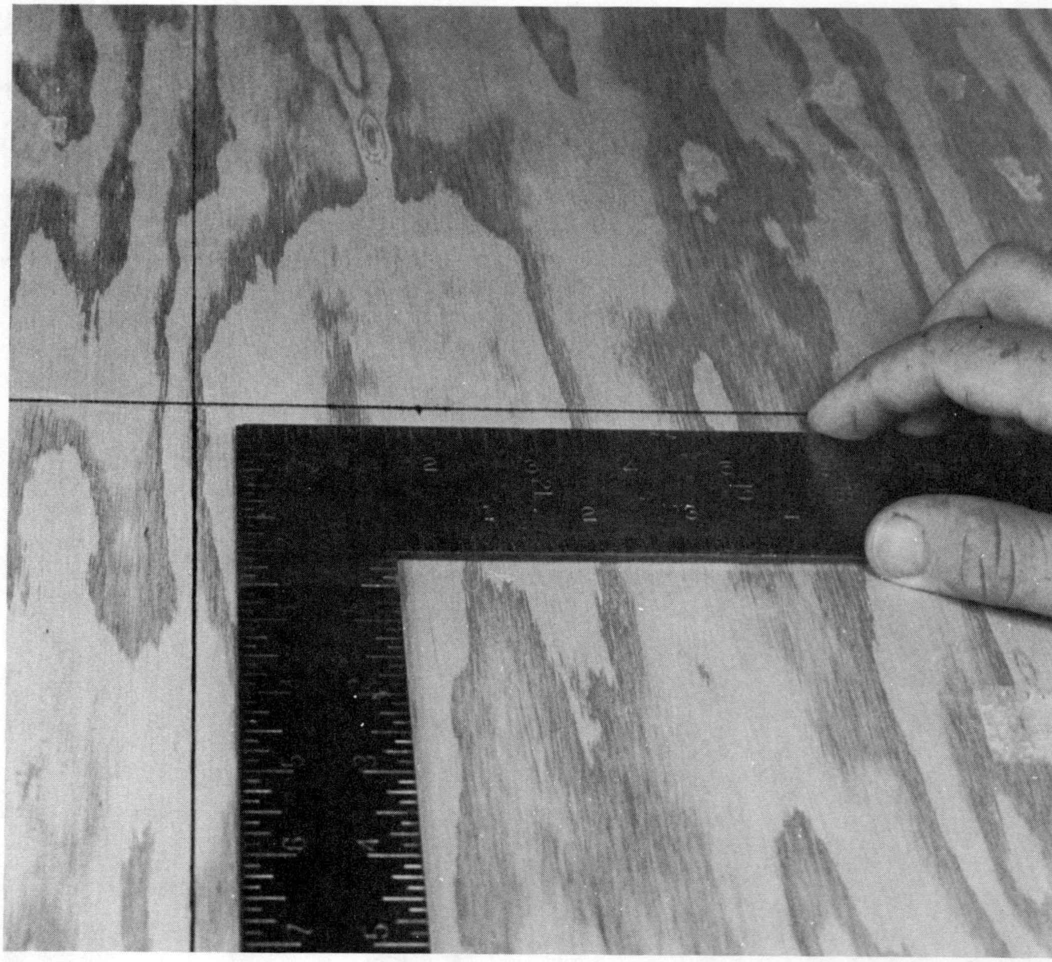

Right:
Place unit or block No. 1 where the lines cross. Then lay units No. 2 and No. 3 on either side of the first unit. Now align the units carefully with Line A and with each other.

Far right:
Unit No. 4 goes directly over unit No. 1. This establishes a pyramid pattern. Continue laying the units around the pyramid.

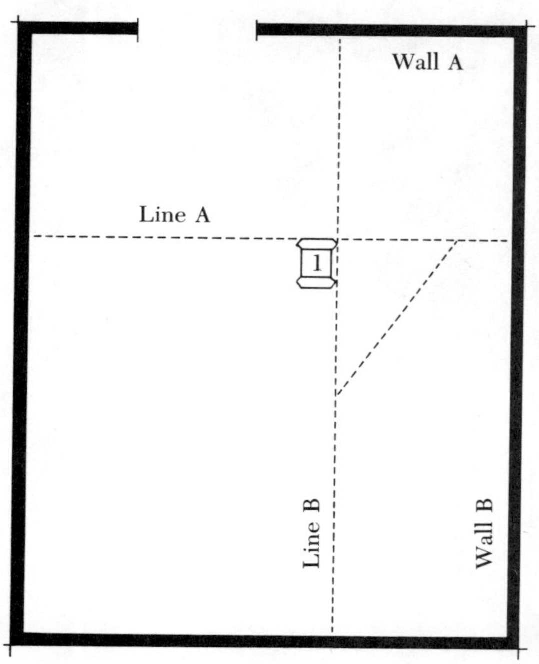

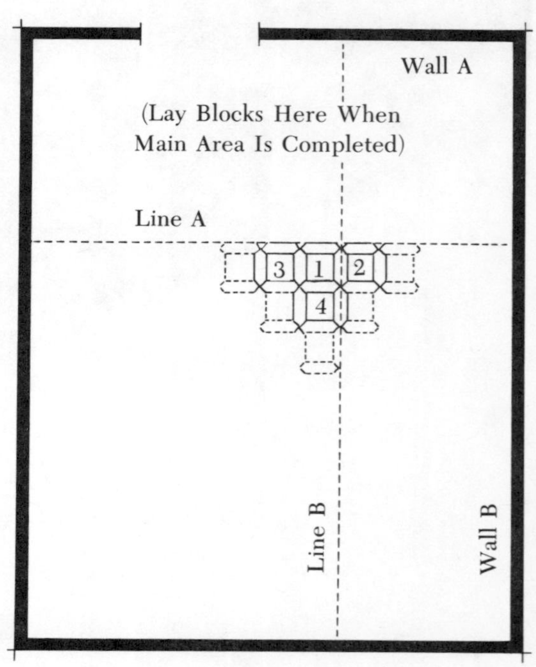

Here is how the pyramid pattern looks. Trim pieces to fit tightly to wall or vertical obstructions. Or, you can face-nail each row to prevent slippage. Let the border units extend as far as possible into door openings.

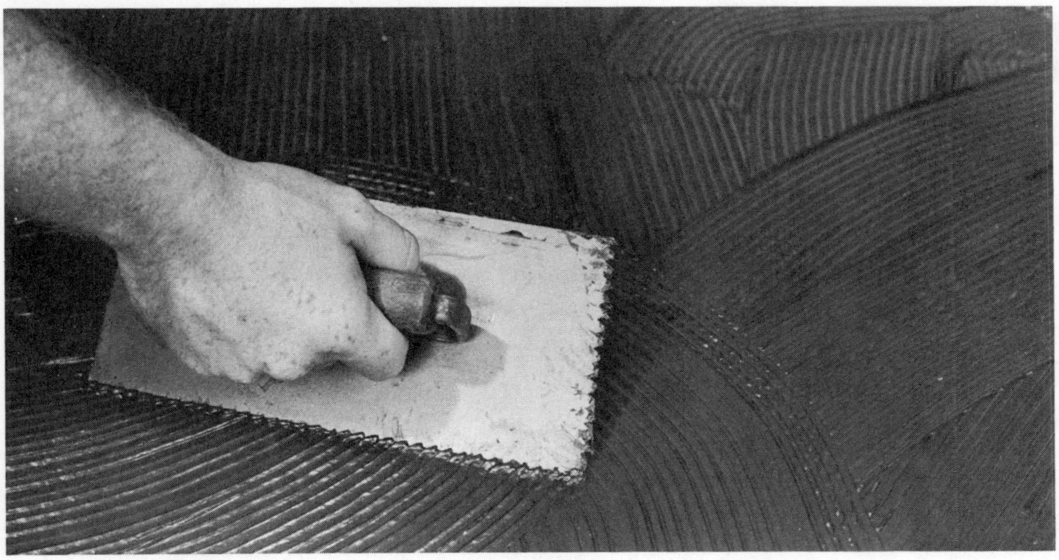

Spread mastic for remaining units with a notched adhesive spreader. Use the type of adhesive recommended by the flooring dealer, and follow the directions on the container carefully. Cover only a small area at a time with adhesive, and make sure it is spread evenly. You will be able to follow the guidelines on the subfloor or underlayment because the notched trowel leaves part of the subfloor bare. Make sure the first 3 or 4 units are properly aligned. The rest of the job will be easy.

Drop the units lightly into position. Do not slide them; sliding will cause the adhesive to collect on the edges, and the units will not fit properly. Parquet, or block, may be scribed to doorjambs or placed under the jambs if they are raised. As you lay the units, put a piece of plywood over them to prevent foot traffic from sliding the units out of position.

Use a buffer block to tap tongue-and-groove blocks tightly together after the units are in position. Tap just enough to assure that the joints are tight. If you smear adhesive on the blocks, wipe off the excess immediately; do not let it dry. When you lay this type of flooring, the room should be 70° or warmer.

Space border tiles about ¼ or ½ inch from walls and vertical obstructions (including doorjambs). This permits expansion and contraction of the wood. Since molding will hide the gap, you don't have to worry about any uneveness or nailheads when you nail the block units into position as suggested above. After the floor is laid, step on every block. This assures contact between the adhesive and the block. Finally, install the molding and the trim pieces. Do not use the room for at least 48 hours so that the adhesive may harden properly.

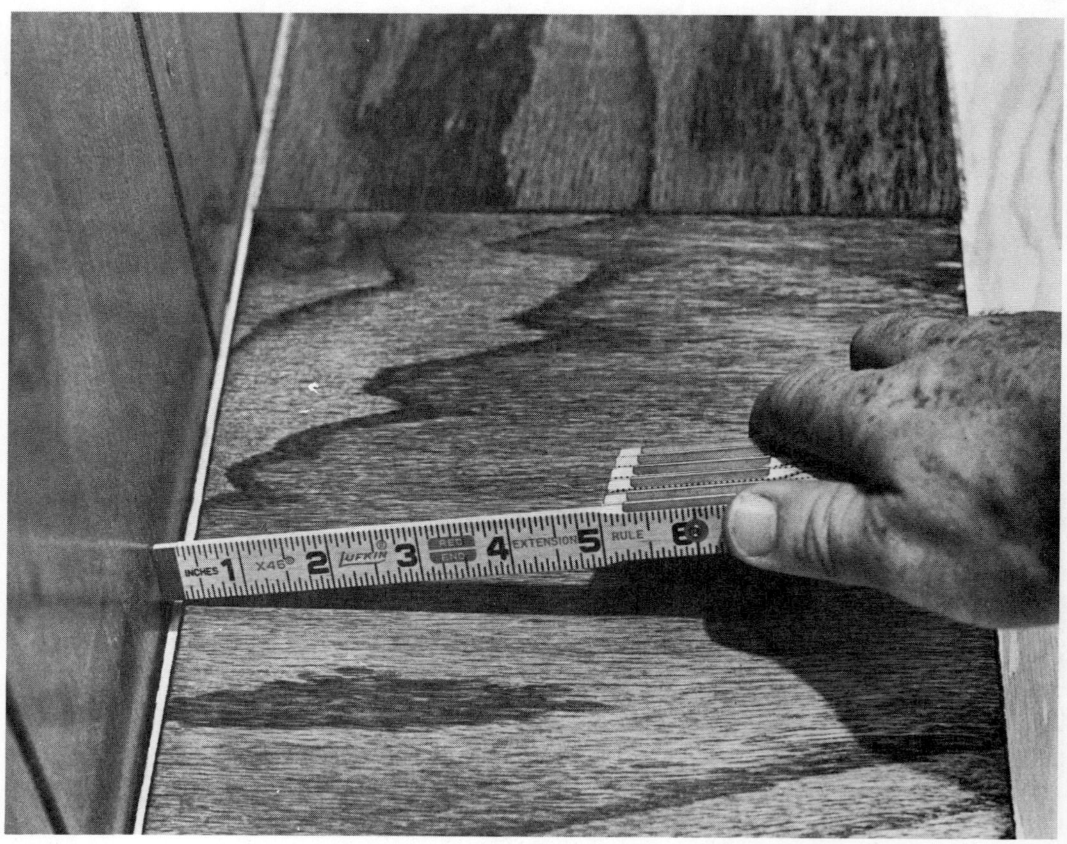

How to install hardwood strip and block floors over concrete

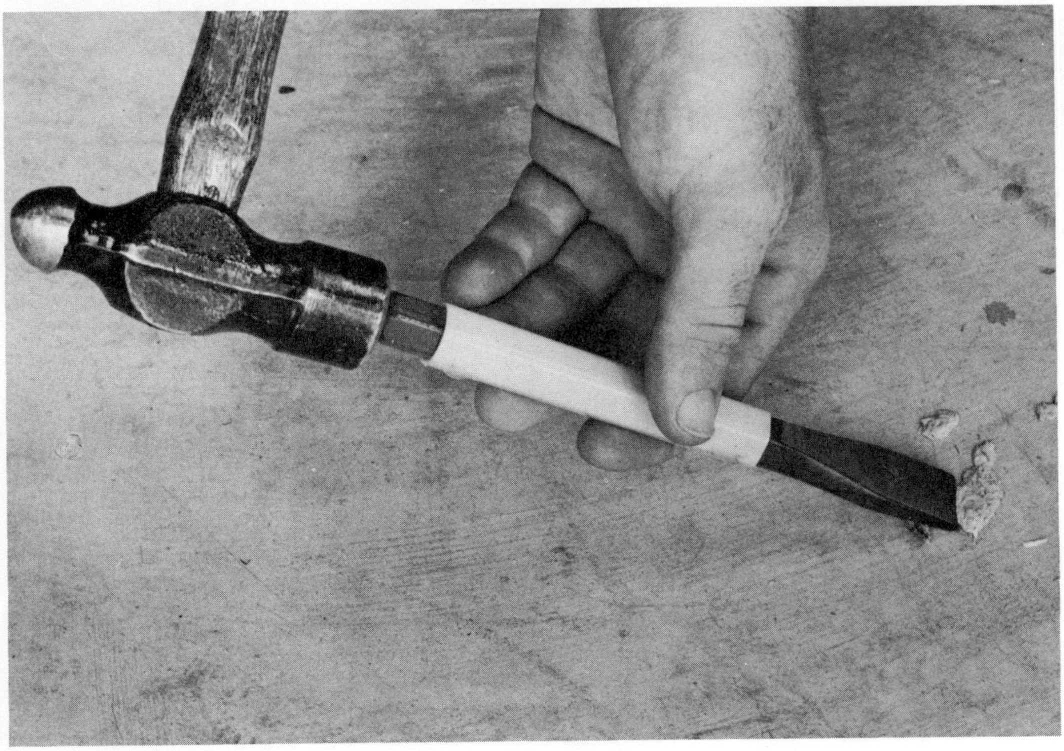

Carefully clean the concrete floor. If daubs of mortar are stuck to the surface of the floor, chip these off with a cold chisel, as shown. The floor must be dry. If there is water seepage anyplace on the floor, find the source and block it. Water is harmful to wood floors.

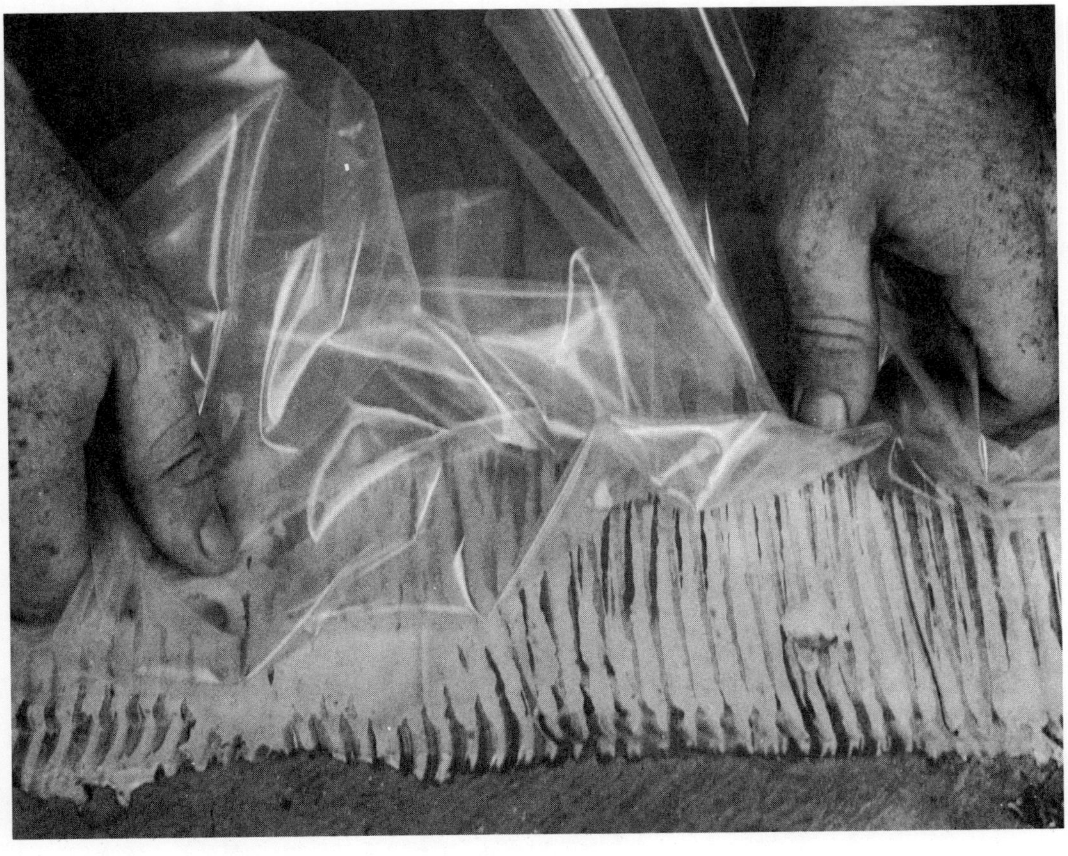

Cover the floor with a special primer for this job. Let the primer dry for 30 minutes; then give the floor a thin coat of mastic. Let this material dry for 30 minutes. Then embed 4-mil polyethylene film in the mastic coating, as shown. Lap the seams 4 inches, and let the polyfilm extend behind the baseboards on all sides of the room.

Use 2 by 4s for screeds (a piece of wood that has a straight or a flat edge). Screeds are sometimes called nailers. The flooring or subfloor will be attached to these screeds. The 2 by 4s should be cut into 18- to 48-inch lengths. Do not use any screeds that are warped. The screeds should be treated with a wood preservative; do not use creosote. They should be staggered in rows on 12-inch centers at right angles to the proposed direction of the finished flooring. The ends should lap 4 inches. Alternate long and short lengths, and leave at least 1 inch between the screed and the baseplates to allow for expansion of the floor. Use a 2 by 6 around the walls of the room, as shown, for a nailing surface.

Set the screeds into position with mastic. You can use cold-stick mastic for this. Let the mastic harden for 30 minutes before you put the screed over the polyfilm. After you go across the room with the screeds, let the mastic set for 24 hours before you proceed with the strip flooring or plywood subfloor.

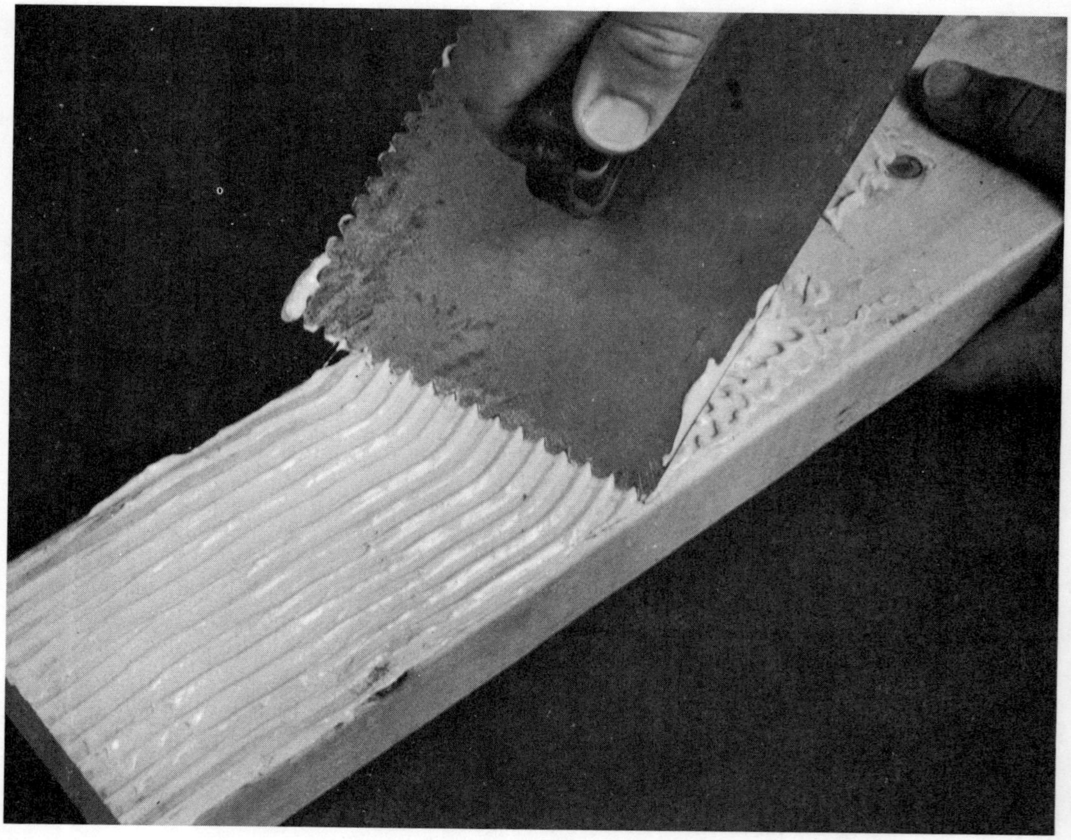

If the floor area to be covered is small, and the concrete surface is rough and uneven, you may be able to use continuous screeds. This technique takes time and patience, but you have the advantage of shimming the screeds level, which means the subflooring will be level. First, cut the screeds to fit the area and counterbore them, as shown, to accept lag bolts, which will be turned down into lead expansion anchors in the floor.

Drill holes in the concrete for the expansion anchors after you have drilled the pilot holes and counterbored the screeds. You will have to position the screeds over the floor so that the measurements will be correct. We recommend that you lay out the area first; then mark the screed and anchor. You can use a star drill to make the holes in the concrete. However, it is easier to use a portable electric drill and a masonry bit to make the holes.

Tap the lead anchor in the hole in the floor, as shown. The top of the anchor must be flush with the concrete surface. The lag bolt then fits through the holes you drilled in the screeds and into the anchor. You will need a set of socket wrenches to drive the lag bolts into the lead anchors.

Use a 2 by 6 nailer along the walls of the room. The screeds butt against this nailer, as shown. The 2 by 4s are laid flat on the floor, which should be covered with 4-mil polyfilm, as explained above.

Lay the screeds on 24-inch centers. If the floor is uneven, you can tap cedar shingle shims under the screeds to level them. The screeds should fit tightly against the floor, and you can accomplish this by loosening or tightening the lag bolts.

The subflooring is nailed to the screeds. Space the nails 6 inches on center. You should use ½-inch-thick (or thicker), exterior grade A/C plywood for the subflooring. After this is down, cover the plywood with polyfilm. This will help to eliminate any moisture that may accumulate under the subflooring. The finish flooring goes over the subflooring; you may use hardwood, resilient flooring, carpeting, or slate.

The hardwood strip flooring in this kitchen highlights the paneling on the walls, the kitchen cabinets, several of the appliances, and the masonry detail. The flooring is laid over a plywood subfloor, and this technique is shown in this chapter. The island range is fastened to the flooring and subfloor with metal angles and by the toenailing method. An area rug in the work area protects this portion of the floor.

Right:
The hardwood strip flooring blends perfectly with the wood of the kitchen pass-through, railings, and furnishings in this home. The construction technique is standard for laying the floor; hardwood strips are placed over a subfloor. The green plants and the area rug contrast with the richness of the wood.

Far right:
Contemporary design is accomplished in this living room with a block floor and rough-textured walls. The blocks are laid in a straight pattern with mastic over a standard subfloor. The pattern design is indicated by arrows on the mastic side of the blocks. An area rug softens the starkness of the room, and the furnishings and painting accentuate the design.

A country or colonial motif is shown with a pegged hardwood strip floor. The pegs are included in the material; you do not have to cut them. Lay the floor as you would lay any other hardwood floor; it is prefinished.

This traditional bedroom is highlighted by a parquet floor. The blocks alternate direction and are attached to the subfloor with mastic. A small rug picks up the colors in the bedspread and skirting, tablecloth, shutters, quilt, and painting. This type of floor is as easy to install as a resilient tile floor. The same layout and application procedures apply, as explained in this chapter.

This formal study has a herringbone design in the hardwood floor, which is set in mastic over a standard subfloor. You may have to do more cutting and fitting of the blocks for this configuration, but you create an attractive pattern. The walls are dark, creating a mellow, restful mood. The oriental rug, pale sofa, and chair coverings lighten the mood of the room.

This formal/informal room with its mixed furnishings has a block floor in a random pattern. We suggest that you lay all individual blocks in the room before you install them over the subfloor with mastic. The blocks may be pulled up and numbered according to position for the final installation. Also, you may put down a quarter of the flooring at a time, spreading the mastic and setting the blocks as you go.

Ceramic Tile and Slate

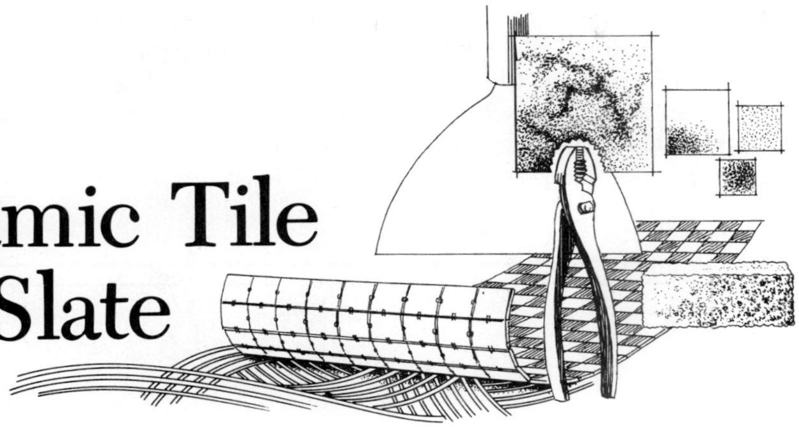

The advantage of using ceramic floor tile and slate is that both products are extremely durable. In fact, ceramic tile and slate will last for hundreds of years. The products will not fade, and they are resistant to moisture and heat.

Ceramic floor tile is recommended for entry halls and bathrooms. It may, of course, be used in any room in your home.

Slate is recommended for entry halls. And, like ceramic tile, it may be used anywhere in your home.

Ceramic tile and slate may be installed over wood subflooring, underlayment, concrete, or vinyl tile. You may not lay ceramic tile over hardwood strip flooring because the moisture from the mastic will make the floor buckle. Also, you should not lay new ceramic tile over old ceramic tile. To remove old tile, follow the directions given in the chapter "Resilient Tile, Sheet Flooring, and Seamless Flooring."

You can purchase ceramic tile in many shapes, sizes, colors, and finishes. If you are an average home-handyman, we recommend that you use the 4- by 4-inch, 6- by 6-inch, or 8- by 8-inch standard floor tiles. They are the easiest sizes to install; you can vary floor patterns, and you can mix or match the tiles to get almost any type of design you want.

Mosaic tiles are made of ceramic. They may be purchased individually or by the sheet. The sheets range in size from 6 by 12 inches to 12 by 24 inches. Some products are bonded to a paper sheet, and the tiles are prespaced. Other products are stuck together with a gumlike substance, and the tiles on these sheets are prespaced. Such tiles are shown in this chapter.

Slate is sold in random colors and sizes. The random sizes can be installed in a flagstone configuration. The square and rectangular products may be installed in a square configuration or in an irregular pattern.

Estimating and measuring for ceramic floor tile and slate

The easiest way to estimate the amount of ceramic or mosaic tile you need for a room is to measure the length and width of the room. Take these measurements to the tile dealer, and he will tell you how much tile you will need for the floor—including tile for normal breakage and waste. The dealer will also estimate how much grout and adhesive you will need.

To lay out the room for ceramic tile and standard slate, you will need to square the room or the area (see drawing).

First, find the center of the entrance into the room. Snap a chalk line from the doorway to the opposite wall of the room. Make sure the line is at right angles to the wall; you can use a square for this. Now, lay a row of tiles along the chalk line from the doorway to the wall. Nail a straightedge against the opposite wall, and butt the row of tiles against the straightedge. Nail a second straightedge at a 90-degree angle to the first straightedge. Use a square to make sure it is a perfect 90-degree angle. The room is now squared.

Start tiling at the squared corner. Since the room is not square to start with, you probably will have to cut at least one row of tiles to fit.

This drawing shows the straightedges that have been measured and set; these guidelines will allow you to lay the tile accurately. Be accurate in these measurements. If the measurement is slightly off, the error will be compounded throughout the room.

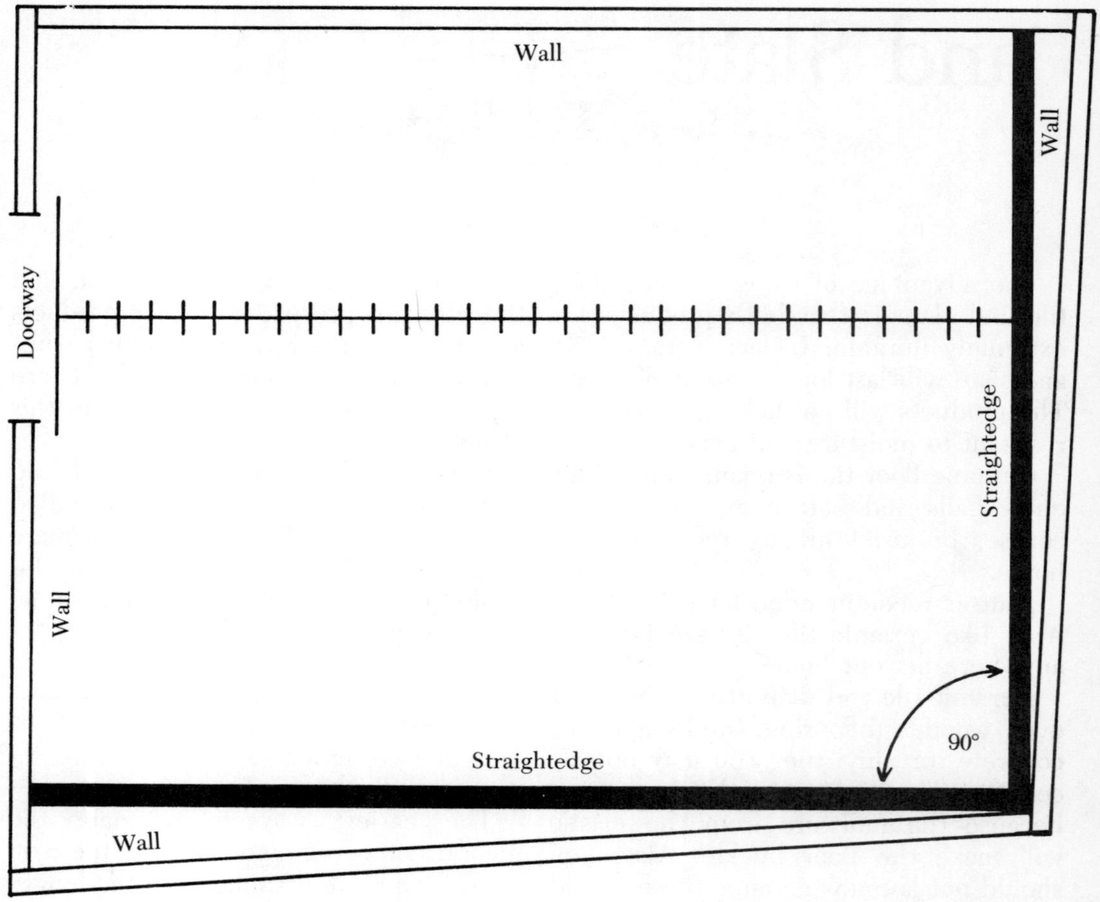

How to set ceramic floor tiles

To install ceramic floor tile, you will need these materials and tools:

- 13-ounce claw hammer
- Carpenter's framing square
- Flexible tape measure
- Chalk line and chalk
- Notched adhesive spreader
- Sponge-surfaced trowel for applying grout (or wooden float, steel trowel, or wide wall scraper)
- Small piece of concrete block
- Tile nippers or pliers (or rent a tile cutter for straight cuts)
- Glass cutter
- Water bucket
- Sponge
- Tile
- Grout
- Adhesive
- Straightedge

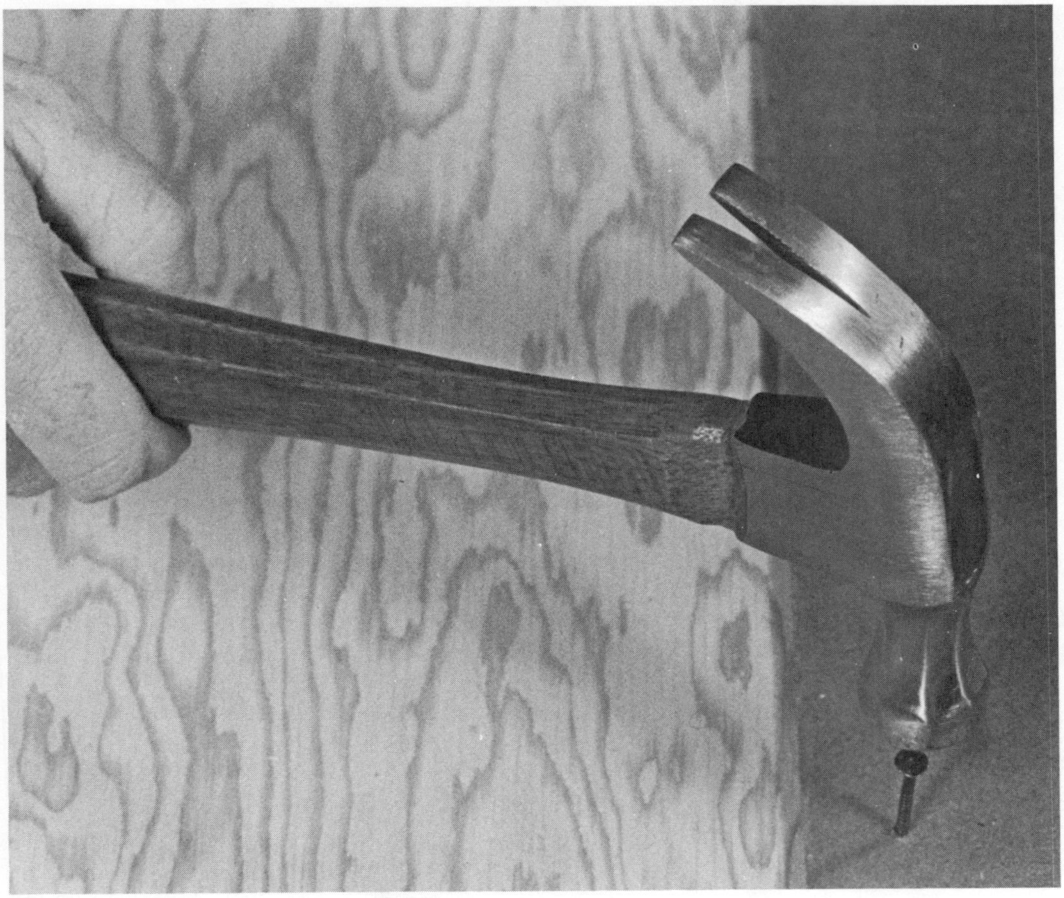

Loose flooring and subflooring must be removed or nailed tightly. If the subflooring is in bad condition, cover it with a tempered hardboard underlayment. Use ¼-inch-thick material, and stagger the joints. Since the thickness of the tile—and underlayment, if you use one—will usually raise the floor from ½ to ¾ inch, the bottoms of doors must be trimmed. Also baseboard trim, heating/cooling registers, and electrical floor switchplates must be removed. Besides hardboard, you can use plywood and particleboard for underlayment. If you use plywood (or particleboard), let this material stay in the room for 72 hours before you install it. The panels have to adapt to the humidity in the room. Also, make sure all wax and loose debris are removed from the floor.

Use ceramic floor tile for floors—not ceramic wall tile. The difference between the two is shown here. Wall tile tends to break under weight; therefore, it is not practical to use on the floor. Ceramic floor tile is usually ⅜ inch thick for 4- by 4- inch tiles and ½ inch thick for 6- by 6-inch tiles. You can buy tiles at some building material outlets, ceramic tile dealers, and general merchandise stores.

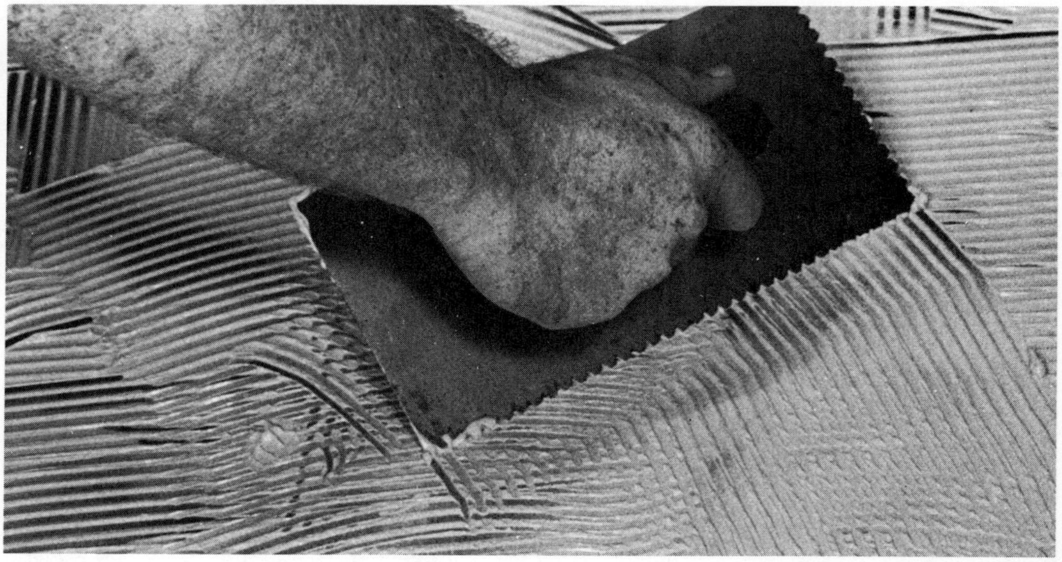

Special mastic is used to set ceramic floor tile. Spread the mastic over the floor with a notched trowel; apply to about 3 square feet at a time, starting at the squared corner. Spread the mastic evenly, and follow the directions on the container for thickness, setting times, and cleanup.

Some ceramic floor tiles have tiny lugs on the edges. These lugs space the tiles properly. If the tiles you buy do not have lugs, make spacers from 1/16-inch-thick cardboard. You will need at least 20 of these spacers so that you can lay several tiles at one time. Put down all the field tile first. Then fill in the border tiles.

To fit tiles at the border and around obstructions, score the face of the tile with a glass cutter that is guided by a straightedge. Use plenty of pressure on the glass cutter; it will lightly score the tile. Once you have used the glass cutter on ceramic tile, it cannot be used to cut glass. Throw away the glass cutter after you finish the tiling job.

Snap the scored tile over a piece of aluminum angle, molding, or a 20d nail. Apply equal pressure on the edges of the tile with your thumbs. If the snapped edges are jagged, you have not applied enough pressure to the glass cutter; discard this tile and try again. Be sure the molding, nail, or aluminum angle is centered on the scored line for an even break. You may also rent a tile cutter for this. A tile cutter is a good investment if you have a lot of tiles to cut and fit—especially if the cuts are less than half the width of the tile.

Smooth rough (not jagged) edges of cut tile by rubbing them across a concrete block, or you can buy an abrasive block made especially for this job. You can also use a brick to smooth the edges. The edges do not have to be perfectly smooth; the grout will cover the joint so that it does not show. Since you will be cutting off the spacing lugs on the tile, you will probably need spacers to maintain the same even grout lines.

To fit tile around pipes and other obstructions, mark the shape of the obstruction on the tile with a wax pencil or a felt-tipped pen. Use a template for this, as explained below. Score the irregular shape, if you can, with the glass cutter. Then, with pliers, nibble away the excess tile. Take tiny bites with the pliers. Otherwise you will break the tile. Test the tile in the space for fit occasionally. When the tile fits properly, stick it into position with a dab of adhesive. You don't have to worry about rough edges; the tile grout will hide them. If you have a lot of obstructions to go around, we recommend that you buy regular tile nippers. The nippers will do a better job than pliers; however, if you only have a few tiles to fit, nippers are not worth the investment.

To fit ceramic tile—or any tile—around an irregular shape, first cut a cardboard template the same size as the tile. With scissors, trim the template to approximate the irregular shape. Test the template in the void, as shown, and continue trimming the template until it fits the shape. Now use the template to mark the proper cut on the tile. Since tile grout will fill the joint, the ceramic does not have to fit perfectly. Other tiles should be cut as perfectly as possible.

When tiles have to be cut to fit against a wall, lay a new tile over the last tile you set, as shown. Make sure the top tile fits perfectly over the tile beneath it.

Place another tile over the second tile. Butt this top tile against the wall. Now, using the top tile as a guideline, mark the cutoff line along the face of the second tile. With the glass cutter, score the second tile along the cutoff line, snap the tile, and smooth the edge. This tile will now fit the void between the last full tile you set and the wall. Score the tile slightly inside the mark to leave space for the grout. You do not have to worry about space next to the wall since the molding will cover any small imperfections.

When all the tiles are in place, let the adhesive set for 24 hours before you grout the joints. Wipe away any excess adhesive as you set the tiles—not after the entire job has been done. The grout should be mixed to the consistency of thick whipped cream. Apply it to the joints in the tile with a rubber squeegee or a sponge-faced trowel. Work the grout into the joints. If you are tiling a small area—such as a small powder room—you can use a wide wall scraper to grout the joints. The trick is to apply plenty of grout. Let the grouting job set for 24 hours. Then check for any voids in the grout. If you spot any voids, fill them with grout.

After you have filled the joints with grout, clean away any excess on the face of the tiles with a medium-wet sponge. Rinse the sponge frequently, and do not apply a lot of pressure on the sponge, or it will damage the fresh grout lines. After the grout has dried, the tiles may have a film on them. If so, you may be able to wash off the haze with water and a sponge. If not, use a commercial tile cleaner, for the specific type of tile you are using, and follow the instructions carefully.

How to set mosaic tiles

Mosaic tiles in sheet form are easier to set, in our opinion, than individual 4- by 4-inch floor tiles or 1- by 1- or 1- by 2-inch individual mosaic tiles. The tile goes down fast because you are working with large sheets, and the tiles are pre-spaced so that you don't have to worry about the spacers.

To set mosaic tiles:

1. Prepare the floor surface. Nail down any loose underlayment or subflooring, and countersink any protruding nailheads in the underlayment of subfloor. Also, remove any debris, loose paint, or loose tiles. The floor should be as smooth as possible.
2. Square the room. Follow the same procedures as described for 4- by 4-inch floor tile. Double-check these measurements. Any error that is made will be compounded.
3. Start the tiling job in the squared corner.

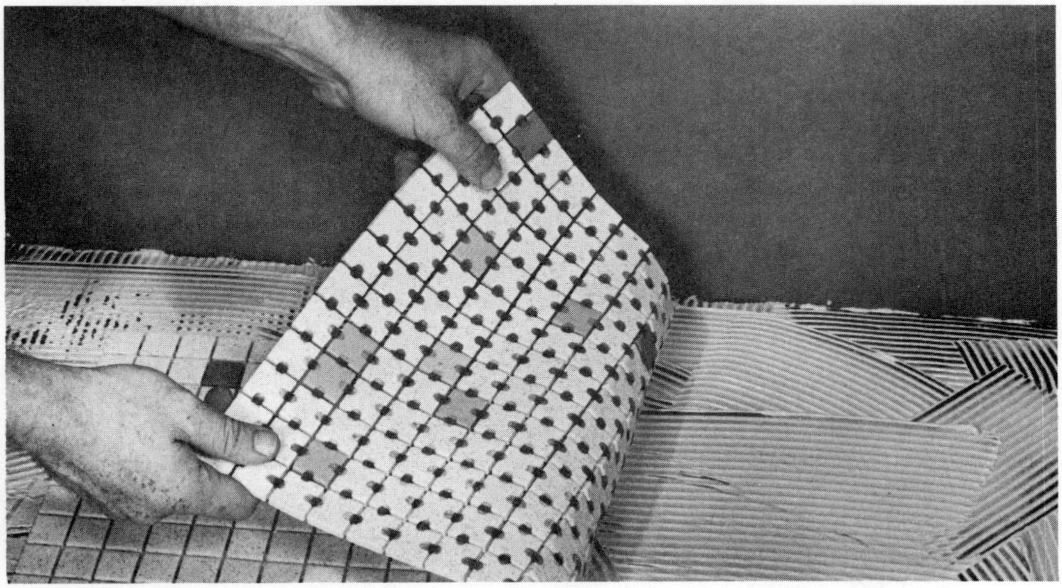

Lay the sheets of tile on the adhesive. Use a notched trowel to apply the adhesive evenly to the floor. Do only a 3- to 4-foot-square area at a time. If the tiles are faced with paper, the paper covering should face upward—toward the ceiling. After the sheets are down, slip them into position, and press the tiles firmly into the adhesive. When the job is finished, wet the paper and peel it off so that you can grout the tiles.

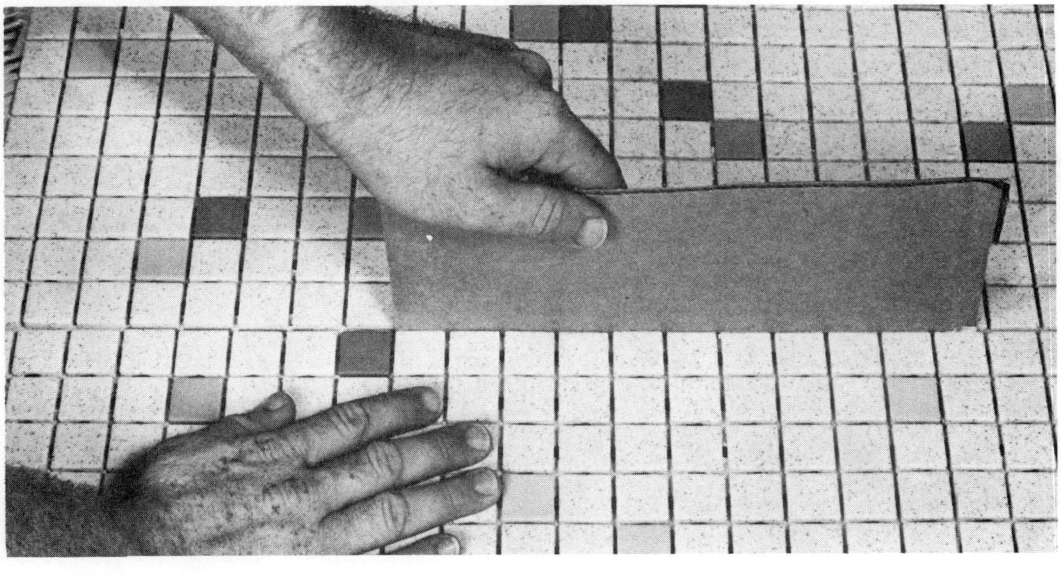

Use a spacer to space the individual sheets of mosaic tile. The thickness of the spacer should be the same width as the joints in the sheet material. You may slide the sheets in the adhesive for a perfect fit. After you have put down four sheets, lay a small piece of plywood or hardboard over the tile surface. Then tap the hardboard lightly with a hammer or a rubber mallet. This will help to set the tiles in the adhesive.

Right:
Use a razor knife to cut the prespaced tiles. Include several extra sheets of tile in your material estimate for trimming and fitting purposes. For fitting next to walls, use the same technique with the sheet tile as you do with individual 4- by 4-inch tiles. Place a full sheet of mosaic tile over the last sheet you set. Then mark the tile for cutting.

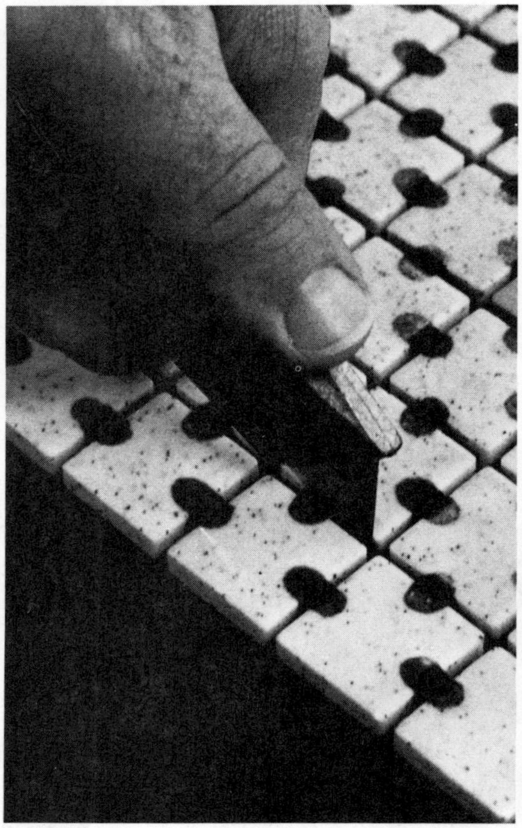

Far right:
The adhesive that holds prespaced tiles together should be cut off the edges of individual tiles for fitting purposes. Use a razor knife for this; it doesn't take a lot of pressure on the knife to remove the adhesive.

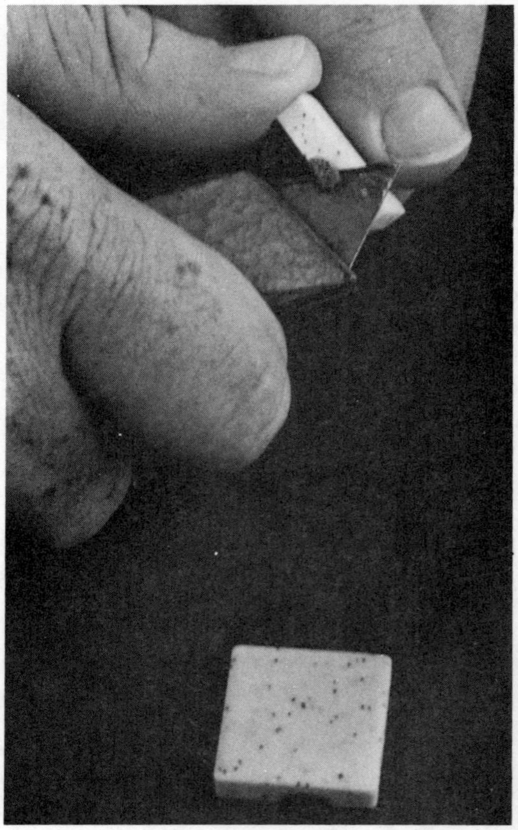

To fit the tiles in small gaps, trim the tiles with pliers, as shown. If you have a lot of tiles to trim, invest in a pair of tile nippers. Take small bites with the pliers or nippers. If you break the tile, start again with a new tile. After the floor has been laid, grout the joints the same way you do with 4- by 4-inch tile, as discussed earlier. If the tile is covered with paper, soak the paper; then peel off the paper. Let the job set for 24 hours before you grout the tiles. After the floor has been grouted and cleaned, wait at least two days before you walk on the floor.

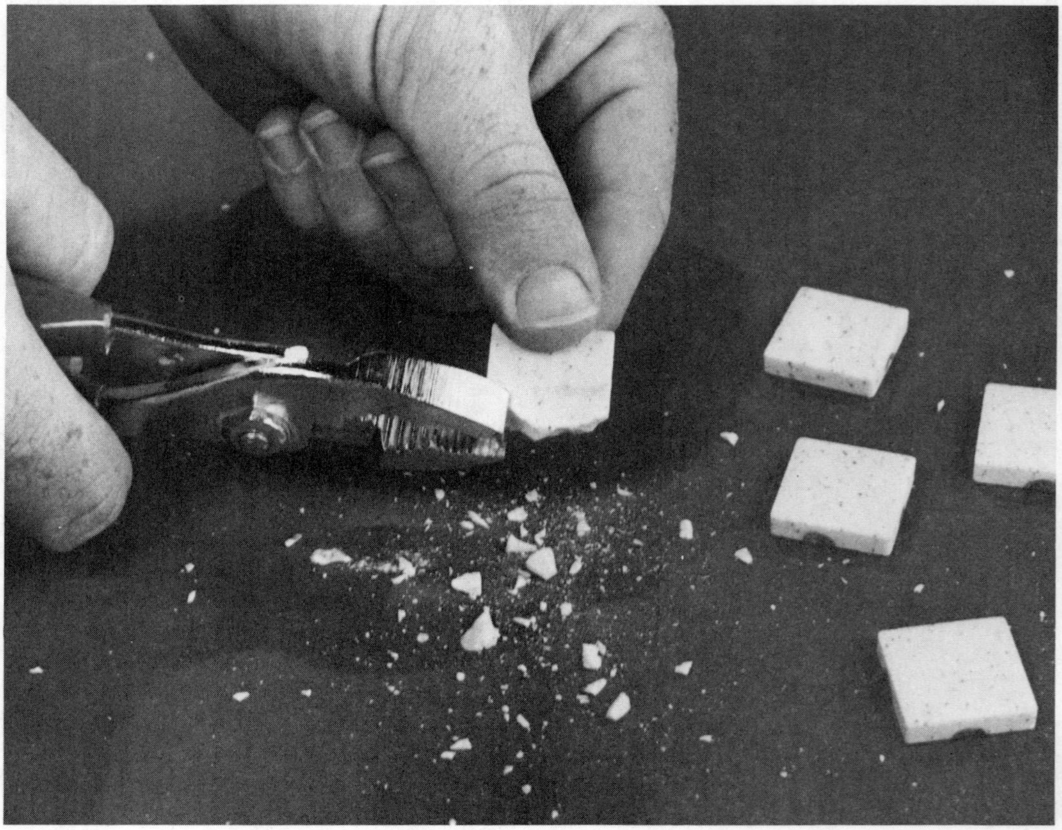

How to lay slate floors

We recommend that you buy packaged slate that is already cut to size and that you then smooth both sides, or buy slate that is smooth on one side and fissured on the other side. Slate from a quarry is rough and must be installed in a bed of concrete. The packaged material can be set with adhesive.

The cost of a slate cutter is prohibitive. However, you may be able to rent a slate cutter from the dealer or a rental outlet. If not, we suggest you lay out the slate floor *dry*, and carefully mark any slate pieces that need cutting. Take the slate to a dealer or a tile contractor and have the slate cut. You cannot cut slate with a ceramic tile cutter.

To install slate you will need these tools and materials:

- 13-ounce claw hammer
- Carpenter's framing square
- Level or straightedge
- Flexible tape measure
- Chalk line and chalk
- Notched adhesive spreader
- Sponge-surfaced trowel (or a steel trowel)
- Water bucket
- Sponge
- Slate
- Grout
- Adhesive
- Slate cutter
- Rubber hammer

Slate may be laid over a concrete, wood, or composition floor. The surface of the floor must be smooth; any loose boards should be tightly nailed, and the surface must be clean. Slate is heavy. If you are going to put down a lot of it and are doubtful about the structural support of the floor, have a professional flooring contractor look at it. For fitting purposes, we recommend that you position the slate *dry* over the floor area. Then mark any pieces that have to be cut. The smooth surface of the slate goes on the floor; the fissured surface goes up. As you remove the pieces after the dry run, number them so that they may be returned to the same position. Then spread the adhesive on the floor with a notched trowel. Put down only a 3-foot-square area at a time. Take a board and a rubber hammer and tap the slate into the adhesive (as shown).

This slate is spaced at intervals of about ⅜ inch. However, this is not a hard and fast rule; you can space the slate almost any width that looks attractive. You will need a spacer block to make the joints uniform, as shown. There are many patterns you can make with slate. Ask a slate dealer to help you.

Keep the slate fairly level. It is important to level slate that has a fissured surface. Check this surface frequently with a straightedge or a level. You can slightly raise or lower the slate with adhesive.

Grout the joints of the slate after the adhesive has set for at least 24 hours. Use a sponge-faced trowel or a steel trowel for this. Make sure the joints are full of grout. You can buy grout in colors to highlight the slate.

Wipe off the excess grout with a damp sponge. Then polish the slate with a dry cloth. Slate does not need much maintenance. When the surface is dirty, simply wipe it with a damp cloth or mop and polish it with a dry cloth.

Quarry tile completes this attractive sunroom. The tiles are laid over mastic, and they are grouted with a cement-based tile grout. The tiles are laid diagonally in the room, and you may have to do a considerable amount of cutting to fit the tiles that butt against the wall. The subtle red clay color of the tile blends well with the furnishings and the lattice design on the walls and ceiling.

The uniformity of the slate floor adds to the architectural lines of the stairway and landing in this entrance. The tiles are 12 by 12 inches square, and they are laid in mastic on a plywood subfloor. The grout is standard white tile grout that is level with the tiles. Level the grout with a sponge after it starts to set—a process that usually takes about an hour. The resilient floor tile in the family room leading to the patio provides continuity to the area and gives the illusion of width to the space, which is long and narrow with a high ceiling.

Multicolored slate flooring is used throughout the foyer of this home and in the adjoining dining room. Slate flooring is laid in mastic, and the joints are grouted. The subfloor must be strong enough to handle the weight of the slate. The joists should be set on 16-inch centers. How to lay a slate floor is explained in this chapter. The slate design is softened by an oriental rug and the upholstery of the dining room chairs. Therefore, the floor is a perfect foil for the rich colors of the furnishings.

Glazed quarry tile complements this foyer. The tiles are laid in mastic over a plywood subfloor. The joints are grouted with regular tile grout; they are slightly concave, which gives an added dimension to the flooring. Accomplish the concave configuration by running your index finger in the joints after the grout has set for about 30 minutes. You can also use a stick for this, but the end of the stick has to be square and beveled at a 45-degree angle. The mullions of the doorway are painted to correspond with the joints in the tile and brick and the color of the walls and ceiling.

Octagonal ceramic tile has a formality to it, but note how well it blends into the informality of this home overlooking a lake. One wall is paneled to accent the tile, and the paneling is extended to the wooden deck and railing. The tile is set in mastic over a sound subfloor, the joints are grouted, and the grout is flush with the tops of the tile. Make sure that the subfloor is of standard construction. Use ¾-inch sheathing over joists spaced 16 inches on center.

Resilient Tile, Sheet Flooring, and Seamless Flooring

Three popular man-made flooring materials are resilient tile, sheet flooring (similar to resilient tile but in sheet form), and seamless flooring, which is a plasticlike covering that resembles sheet flooring.

The three types of flooring described in this chapter may be used satisfactorily; we have no recommendations as to type. The prices are competitive.

However, we recommend that you look at the three types and let the flooring dealer help you pick a product that suits your needs. There are hundreds of resilient tile and sheet flooring patterns. Seamless flooring utilizes a chip system, and the chips may be turquoise, tan bark, flamingo, avocado, sandstone, sand, or alabaster.

Regardless of the product you purchase, be sure to ask the dealer for any manufacturer's literature about the product. It can be helpful when used in conjunction with this book.

Preparing the subfloor

If you have to remove old flooring—such as resilient tile or ceramic tile—use a flat steel hoe or ice scraper—available at most hardware and home center stores. Scrape the old material off the subfloor. Before the new floor can be installed, you will have to remove any mastic and debris from the subfloor. If most of the floorcovering on the old floor is tight, you do not have to remove all the old floor before installing the new one.

If the floor needs an underlayment, use the tempered hardboard or exterior grade plywood as mentioned elsewhere in this book. Stagger the underlayment joints. Four corners should not meet at the same point. Use ring-type nails or cement-coated nails to fasten the underlayment to the subflooring. Make sure all nailheads are driven flush with the underlayment surface. Store the underlayment in the room in which it will be used for 72 hours before installing it.

Resilient tile flooring. Resilient tile flooring is easy to install. You can buy it with or without adhesive backing. Installation basics for self-adhesive flooring are explained elsewhere in this chapter. The same techniques may be applied for installing floors with an adhesive and a notched adhesive spreader.

Before putting down a new resilient floor, make sure the subsurface is smooth, clean, and dry. If the floor is painted, the paint must not be flaking, peeling, or cracking. If you are installing new resilient tile over an old floorcovering, the old covering must be tight on the subfloor.

You may install new resilient tile flooring over any old floor except asphalt or vinyl tile. If the floor is wood, replace any damaged boards and countersink any nail pops. If the floor is in poor condition, you may have to cover it with a plywood underlayment. Do not use a particleboard underlayment.

Sheet flooring. This material is manufactured in huge rolls. It is easy to install; fasten it down with staples around the edges of a room, or nail it to the subfloor. It is so flexible that it can even be folded like a blanket. The material also has a built-in margin for error. You can trim it imperfectly and still obtain a perfect fit, because the material will stretch somewhat to make up for any cutting errors.

Seamless flooring. This material is in liquid form. You simply paint it onto a floor surface, sprinkle on the decorative chips, and seal the chips. Seamless flooring may also be used on countertops, walls, bath and shower enclosures, and patios. It is dirt resistant, weather resistant, stain resistant; moreover, it will not slip, will not yellow, and is easy to care for. Also, the material may be applied over old floors, subfloors, and underlayments that are not in perfect condition.

Choose the flooring you prefer; then, let the flooring dealer estimate the amount of flooring and adhesive you need—if adhesive is used. He needs only the measurement of the room.

If you decide to install a seamless floor, you will have to install an underlayment over tongue-and-groove wooden strip floors as well as parquet floors, cork or rubber tiles, linoleum, vinyl, vinyl-asbestos and asphalt tile that is badly worn, particleboard, and flakeboard. You cannot apply seamless flooring to concrete that is in poor condition, ceramic or glazed tiles, metal, terrazzo, damp surfaces, and wood in exterior locations.

You will also need these tools for the respective floors:

Resilient tile. 13-ounce claw hammer, chalk line, chalk, razor knife, notched adhesive spreader (if the tile is not self-sticking), nail set

Sheet flooring. 13-ounce claw hammer, nail set, flexible measuring tape, heavy shears, stapler and staples, razor knife

Seamless flooring. 13-ounce claw hammer, nail set, paint roller, roller cover, paint tray, whisk broom, sanding block

How to install resilient tile

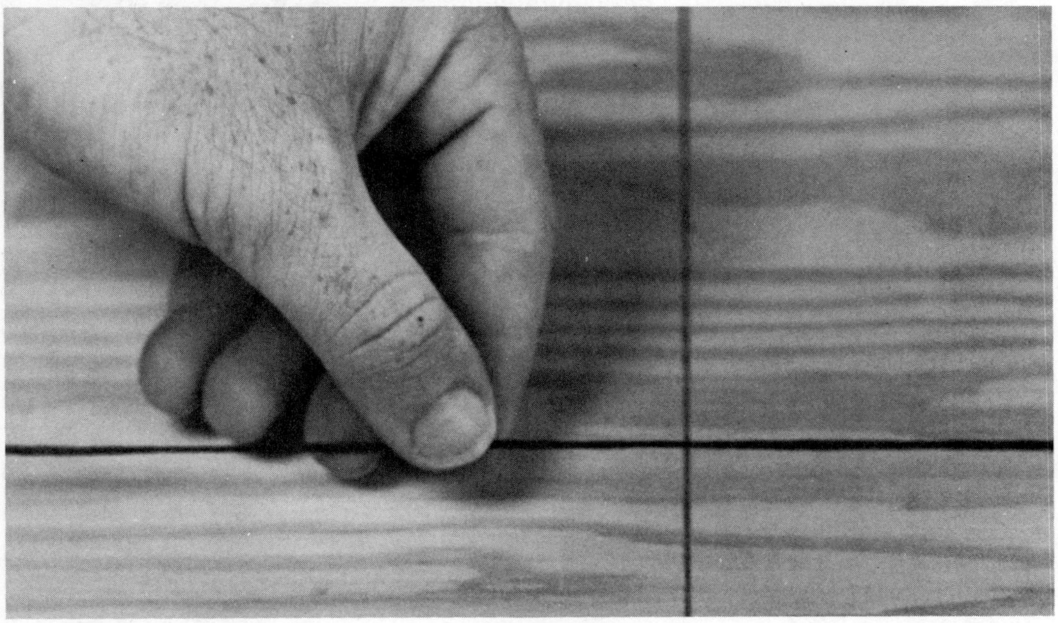

When you install resilient tile, first remove the baseboard from the room; use an old chisel to pry it off. Divide the room into quarters, and snap a chalk line at the halfway point between the long and short walls. These lines will mark the center of the room, and this is where you will start laying the tile. Do not guess at the center point. Double-check the chalk lines before you continue the project. You can do this with a carpenter's square where the lines cross. If the lines are not accurate, snap them again. You must start on the square.

Check your measurements with tile. Lay one row of tile without adhesive along one chalk line and another row of tile along the other line. You may do just a quarter of the room this way, but we recommend that you go completely across the room in both directions. This will give you an idea of how the tile will be spaced. It will also give you the opportunity to change the pattern and measurements.

If the last tile in the row at the wall is less than half a tile width from the wall, move the center line of the room over half a tile width. Snap a new chalk line. Ideally, the space between the wall and the last tile you set should be a full- or a half-tile width. By moving the center line to accomplish this, you will have even rows of tiles on the opposite sides of the room. Since most rooms are not absolutely square, this trick will also eliminate a lot of cutting and fitting.

Self-sticking tile has a paper or a plastic film backing. Simply peel off this backing, and the tile is ready to press into position. If you do not use self-sticking tile, you will need adhesive and a notched trowel to spread the adhesive. Use adhesive recommended by the tile manufacturer. Spread the adhesive in only a small area at a time. When the tiles are set in this area, wipe up any excess adhesive from the face of the tiles and proceed to the next area.

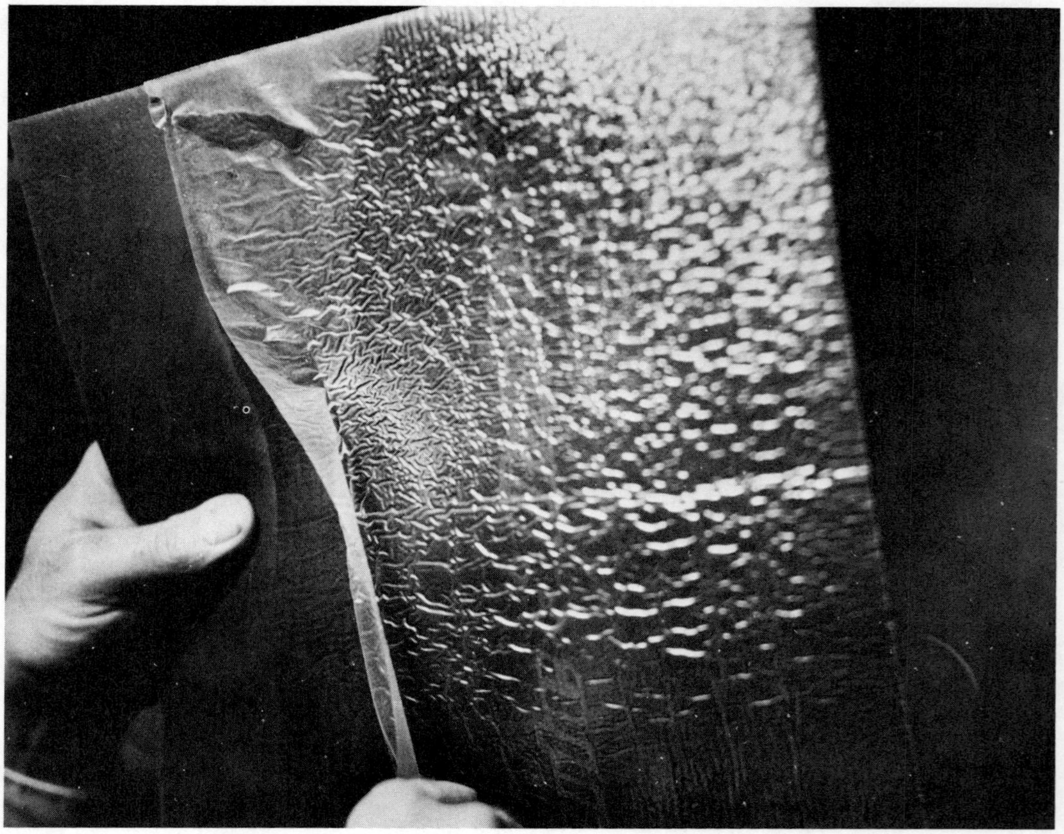

Start tiling at the center of the room where the lines meet. Be very careful to set the first tile accurately; the rest of the tiles in the room will be laid from this center tile. If you are using self-adhesive tiles, butt each tile tightly against the next tile. Be sure the pattern in the tile matches. Once you smooth a self-sticking tile, it is extremely difficult to remove it without breaking the tile. Tile one quarter of the room at a time. Do not trim the border tiles until all the field tiles are in position.

At the edge of the room, lay a whole tile directly over the last whole tile you set. Make sure the edges of both tiles match perfectly. Now place a second tile over the whole tile below it, as shown here, so that the tile butts against the wall or vertical surface. The edges of this top tile must be in perfect alignment with the tile below it. Take your time and make the pattern and edges match.

Using the edge of the top tile as a straightedge, carefully draw a line on the whole tile beneath the top tile. Then cut the marked tile along the scribed mark. The piece you cut should fit the gap between the last whole tile you set and the wall. If you have to go around an obstruction on the floor, make a cardboard template the same size as the tile, marking the cutout configuration. Transfer this configuration to a whole tile and make the cut with a razor knife or heavy-duty shears. The procedure for using a template is shown in the chapter "Ceramic Tile and Slate."

How to install sheet flooring

Since vinyl sheet flooring is available in widths up to 12 feet, you can put it down and not seam the material. The 12-foot sheets are pliable; they may be folded as you would fold a blanket. The first step to laying this flooring is to remove the baseboard or trim pieces at the wall and floor. Make sure the floor is smooth, clean, and dry. Measure the width and length of the room. Double-check these measurements since they have to be transferred to the new vinyl sheet. In another room, unroll the vinyl sheet, and let it reach room temperature. Then transfer the measurements to the sheeting.

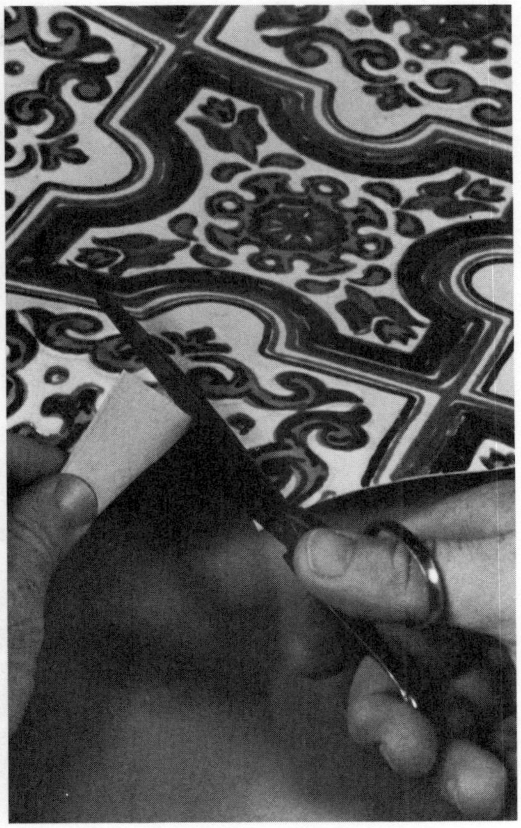

Far left:
Make sure one edge of the sheet is straight, and mark on it as many cutouts as you can. In tight quarters, you may have to cut a cardboard template for irregular trim, pipes, and ducts. If so, transfer the measurements from the template to the vinyl sheeting, as shown.

Left:
Cut the sheeting with heavy shears. Since it is already padded or cushioned, you do not need any special floor treatment. After you have cut the sheet to size, roll it up and install it. Start laying the sheeting at the longest wall in the room. Butt the straight edge of the sheeting against the wall; then unroll the sheeting. The excess will roll up against the other walls in the room; press the sheeting into position.

Trim excess sheeting with a razor knife or shears. At base cabinets and along walls where molding will go, leave about ⅛ inch of excess for expansion and contraction of the material. At doorways, you should protect the edge of the sheeting with a metal strip which screws to the floor or subflooring. You can nail or staple the edges of the sheeting. Trim moldings complete the job, as shown. To allow space for expansion and contraction, slip a piece of fairly thin cardboard between the bottom of the trim and the top of the floor surface. After the molding has been fastened with 4d finishing nails or adhesive, remove the cardboard. If you use nails, countersink the nailheads, fill the holes, and paint or stain the trim.

How to install seamless flooring

The floor has to be clean for seamless flooring. Remove all dirt, wax, oil, grease, and surface coatings, including paint. You may need an underlayment, as explained earlier in this chapter. Apply the base coat of seamless flooring with a brush or a paint roller. Do about 16 square feet at one time. When you are finished with a 16-square-foot area, spread on a full wet coat of laminating plastic.

Evenly spread the color flakes over the laminating plastic. Do this immediately after you apply the plastic; do not let the plastic harden.

Spread the base coat over the remainder of the floor in 16-square-foot sections. Then coat the area with laminating plastic and spread the color flakes over the plastic. Repeat this process until the floor is completely covered with base coat, laminating plastic, and color chips.

With a whisk broom, go over the floor and remove any excess color flakes. Make sure all loose flakes are removed before you continue the job.

Cover the color flakes with another coat of laminated plastic. The flakes are now sealed. Let the floor dry for a day or so. If the floor you are covering is a large one that requires several boxes of color flakes, mix all the flakes together to get an even mix. Also, apply enough chips to the laminating plastic to avoid lap marks. When the floor is finished, it should have a uniform appearance.

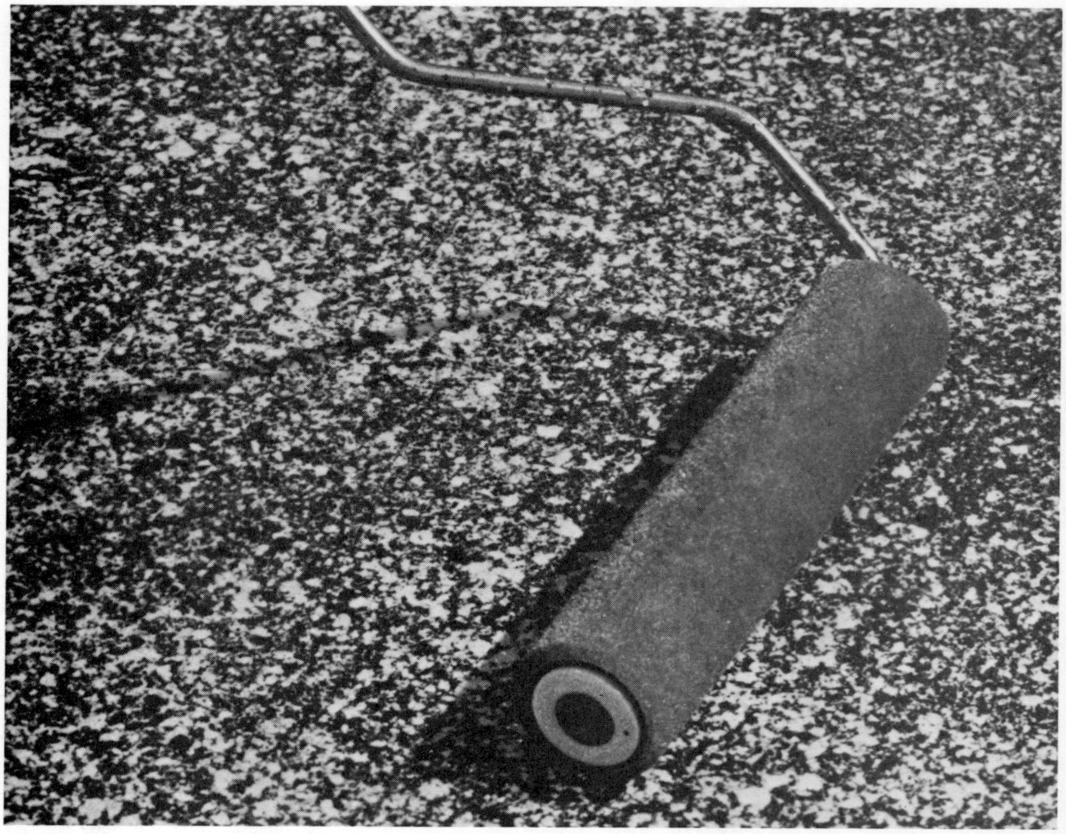

Lightly sand the floor after the laminating plastic is dry. This removes any surface roughness. Wait 24 to 48 hours before you replace the furniture. Depending on the temperature, humidity, and type of surface, it may take several days for the coating to obtain maximum hardness. You can clean tools used in the clear plastic with lacquer thinner.

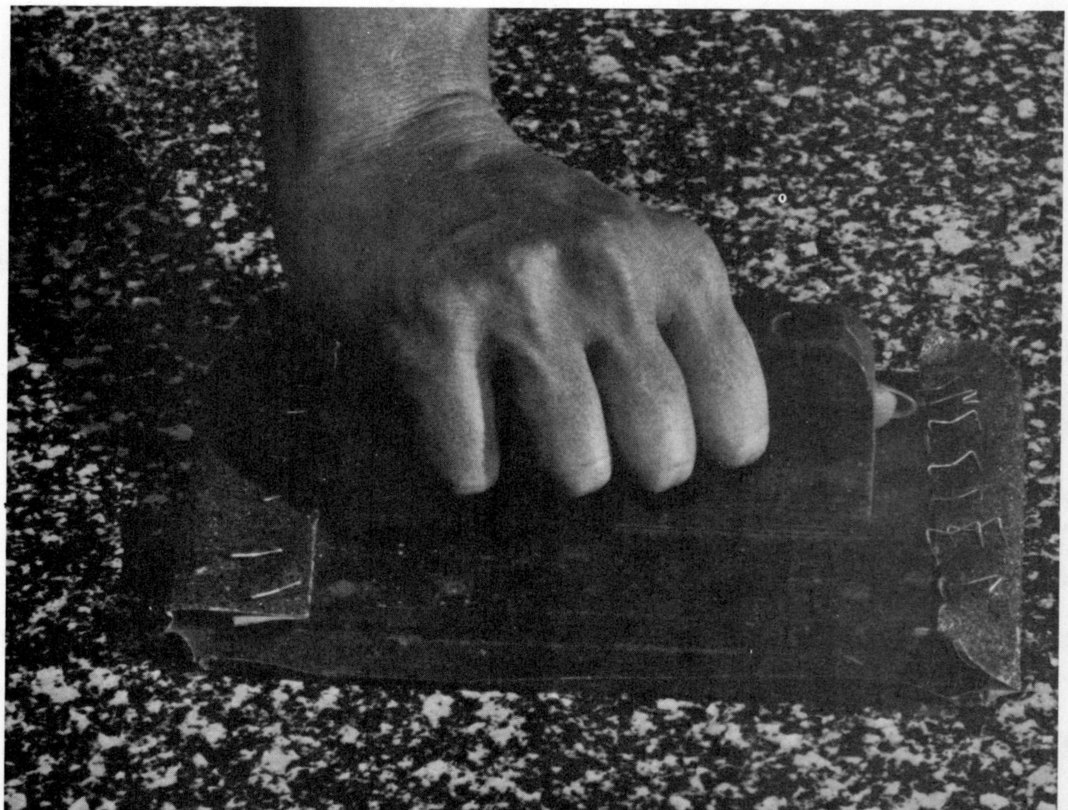

Sheet flooring in a Mediterranean design highlights this kitchen. The flooring may be folded and laid like a blanket, using staples and/or double-faced tape to fasten it down. This procedure is explained in this chapter. The design helps carry out the formal theme established by the well-tailored cabinets and louvered doors that hide the washer and dryer. The wallcovering picks up the color of the flooring, and the colors blend with the flooring and the wood in the cabinets.

Resilient flooring in a simulated brick pattern is extended from the kitchen to the dining room in this home. Since this material is fairly rigid, you cannot bend and fold it. Therefore, you have to measure carefully so that the material fits the area you are covering. Resilient flooring should be installed over flooring that is smooth and tight. See this chapter for installation details.

Carpeting and Carpet Tile

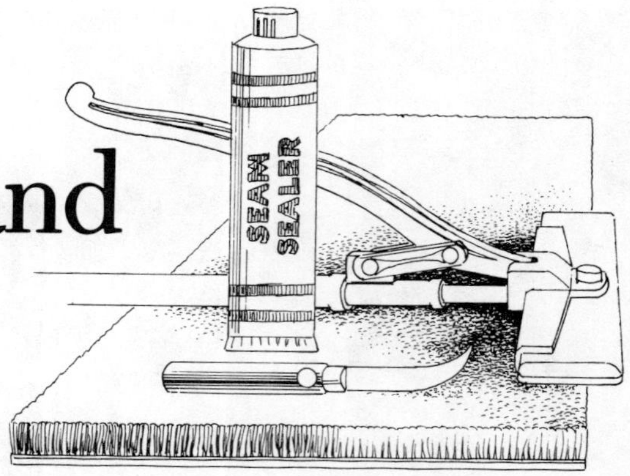

A big selection of carpeting and carpet tile awaits you at a floorcovering dealer. The types and patterns offered today are so numerous that it would be impossible to list them all in this book.

However, we are going to give you several recommendations about carpeting —and carpet tile—and these suggestions, we hope, will help you make a decision.

There are two types of carpeting products available:

1. Cushion-backed carpeting or carpeting that is padded with foam.
2. Indoor/outdoor carpeting that may be used in bathrooms, kitchens, on the patio, or in a foyer—as well as in a living room, dining room, family room, or basement.

Carpeting that is cushion- or foam-backed may be purchased in tile, strip, and roll form. Indoor/outdoor carpeting is available in rolls and tiles.

Cushion-backed and foam-padded carpeting is an interior product. Indoor/outdoor carpeting is designed to be used either inside your home or outside your home.

The price of some carpeting that is sold today includes the installation, as we pointed out earlier in this book. And, frankly, it does take special tools and skills to install carpeting properly. Since carpeting is expensive, you may want to consider some of the problems involved in installing it before you make the decision to do the job yourself. In this chapter, we show the installation techniques involved. Whether you do the installation yourself or have a professional lay the carpeting, you will find it helpful to have an understanding of the methods.

Carpeting terms

Plush carpeting. This product has a surface that is smooth and even. The carpeting has a soft look to it and appears somewhat shaded after it has been vacuumed. This is because the pile of the carpet is closely woven together. The carpeting material stands up straight and is difficult to crush.

Shag carpeting. You can buy long and short shag carpeting; you also have a choice of solid colors, tweed designs, and multicolored yarns. Shag carpeting has a casual look. You would probably not use this product in a formal room setting.

Sculptured pile carpeting. Different lengths of pile are utilized in the manufacture of this product to produce varied surface levels throughout the room. If the pile is uncut, it is called looped carpeting. If the pile is cut and also has multiloops, it is called random sheared carpeting.

Level loop carpeting. This product is exactly what its name implies: it consists of level loops. The loops are tightly constructed; the product is usually ideal for commercial installation because it wears well and is easy to maintain.

Indoor/outdoor carpeting. You don't need a cushion with this product; the carpeting has a latex or a foam-rubber backing that is bonded to the carpeting. This is an excellent do-it-yourself product, and it is fairly inexpensive.

Carpet tile. Carpet tiles in 12- and 18-inch squares are available in many patterns and colors at most carpeting dealers. The advantage of using tiles is that you can quickly and easily change them if several become worn. Also, installation of these tiles is a simple process. Carpet tiles are easier to install than resilient tiles.

Buying carpeting

When purchasing carpeting, there is one rule to follow: buy quality. Quality carpeting costs more, but it will last for years.

Here are some points you need to consider when buying carpeting:

- The more rows of yarn per inch the carpeting has, the closer the weave of the carpeting. Carpeting that is closely woven is usually good carpet.
- Good carpeting has a balanced combination of yarn rows and stitches. If the manufacturer has skimped on yarn—or stitches—the carpeting will be uneven.
- Carpeting with a high pile feels soft under your feet. Carpeting with a high pile also costs a lot more than a medium or a low pile carpet. Inexpensive carpeting sometimes uses a high pile with a reduced number of rows and stitches.
- Carpeting tends to fuzz and shed. However, quality carpeting only does this for a short time. A quality carpet with lots of yarn in it will shed and fuzz more at the outset than an inexpensive carpet.
- The weight of the yarn used in carpeting is important. Always compare the weight of the yarn in different carpeting—expensive and inexpensive. This will help you make a decision. However, you cannot compare different types of yarn fibers—wool with polyester, for example. The weight may be different.
- Jute-backed carpeting is generally used in dining rooms, bedrooms, living rooms, and other low-traffic areas. Rubber-backed carpeting is generally used in kitchens, bathrooms, family rooms, and game rooms, where traffic is heavy. The rubber forms a moisture barrier and makes carpeting easy to clean and maintain.
- Strip carpeting may be purchased in 3- by 5-foot sheets. Most strip carpeting has a self-stick backing.
- Roll carpeting measures from 3 feet to 12 feet in width. You may be able to buy the material in 15-foot widths. The secret is to buy the material as wide as possible—to fit the room, of course—in order to eliminate seaming.

Carpet Fiber Chart

	Acrylic	*Wool*	*Nylon*	*Polypropylene*	*Polyester*
Resilience	Medium	High	High	Medium	Medium
Fade resistance	High	Medium	Medium	High	High
Ease of stain removal	High	Medium	Medium	High	High
Static shock	Fair	High in low humidity	High	Low	High in low humidity
Cleaning	Medium	High	High	High	Medium

Nail down any loose subflooring before you lay carpeting or carpet tile. Also, remove baseboards and trim from around the room. If there are high and low spots in the room, sand them smooth and level. If there are gaps in the subfloor or in the floor over which you are installing the carpeting or carpet tiles, fill them with latex-based concrete (if the floor is concrete) or wood filler. Remove wax, grease, dirt, grime, and other debris from the floor. If the floor surface is concrete, make sure the floor is level, fill the gaps, and block any moisture. Also, remove any oil, dirt, or grease. We recommend that you seal a concrete floor with a paint-on liquid.

Install the nailing strips for the carpeting around the room. These strips are lengths of ¼-inch-thick plywood with tacks protruding through the surface. The strips are face-nailed to the floor around the edges of the room and around obstructions. You can buy these strips at a flooring dealer. Or you can buy a sheet of plywood and make your own strips.

Nailing strips are installed at every point where the carpeting will be stretched and fastened to the floor. To estimate carpeting material, measure the room, including dimensions of door openings and any obstructions. Take these measurements to the flooring dealer; he will figure out the amount of material you will need. If you are installing carpeting, draw the floor measurements on graph paper. This will help you to determine the best width of carpeting to buy and the best locations for seams in the carpeting—if seaming is needed. The time you take figuring out this information can save you money.

Nailers go on stairsteps this way, if steps are to be carpeted. Nail the strips on the lower portion of the risers and on the back edge of the treads. Also, fill in any gaps at the vertical surfaces (walls) so that the carpeting may be fastened at these points. Don't skimp with the nail strips.

How to install carpeting

The tools you will need to install carpeting and carpet tile depend on the type of carpeting you purchase. Here is a list; you will have to determine what you need for the job at hand:

- 13-ounce claw hammer (to nail down loose subflooring)
- Nail set
- Straightedge (use a piece of aluminum angle)
- Razor knife (for cutting both carpeting and carpet tiles)
- Flexible measuring tape
- Chalk and chalk line
- Hot plate for seaming
- Steel trowel
- Carpet stretcher
- Seaming kit (for some carpeting installations)
- Old chisel (for removing molding and trim in the room)
- Stapler and staples (for carpeting and padding, if you use padding)
- Heavy-duty shears (for cutting carpet tile)
- Nails (for nailing loose subflooring)
- Medium-grit abrasive and sanding block
- Wax and grease remover

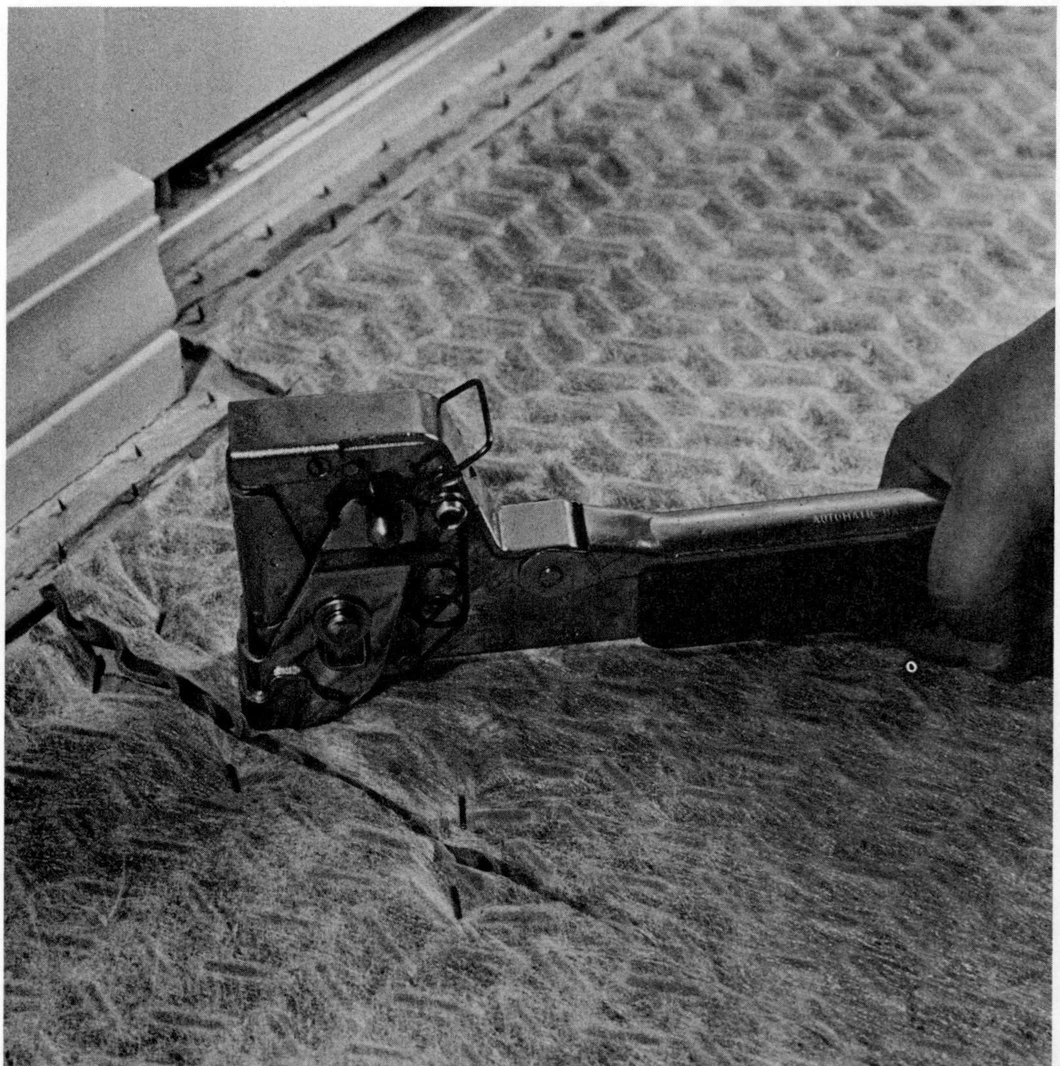

Carpet padding is fastened to the floor with staples. There are many types of carpet padding. We suggest that you rely on your flooring dealer to choose the best type of padding for the carpeting you buy. The trick to stapling the padding to the floor is to space the staples at 4-inch intervals. Don't skimp on the staples, but you don't have to overdo them either. Double-faced tape may be used to stick some carpeting to the floor, eliminating the nail strips. Your flooring dealer will help you decide which technique you should use.

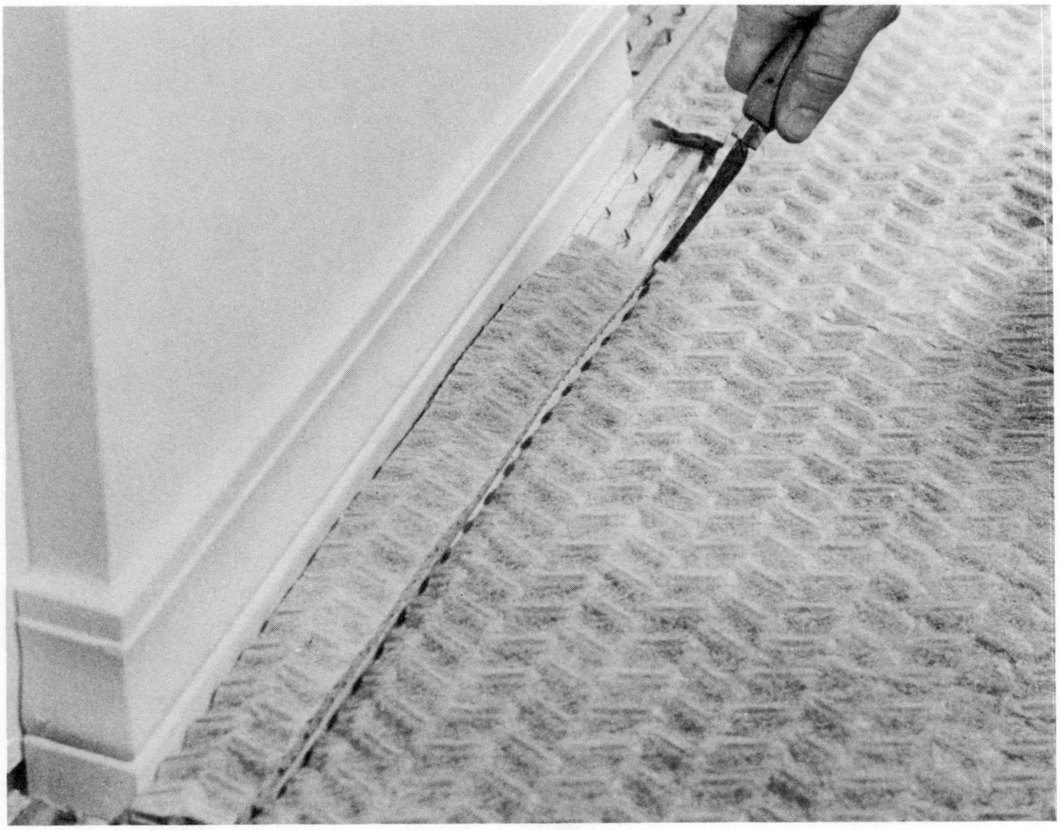

Trim the padding with a razor knife so that it does not cover the nail strips. Be as accurate as you can; the padding must be smooth. Before you lay the padding, unroll it in the room and adjust it for fit. This can save time.

Unroll the carpeting in the room in which it will be installed. You should have at least a 2-inch overlap at all edges in the room. Also, be sure that the pattern of the material and the nap of it matches all the pieces you will be fitting. To go around obstructions within the room, lay the carpeting up to the obstruction and cut the carpet to fit. This cut does not have to be exact since you can trim away any excess carpeting after it is butted against the obstruction. However, cut as accurately as you can. Press down the carpeting to remove any air bubbles under it.

A cut around an obstruction looks like this. Another cut will be made on the far side of the obstruction so that the carpeting will fold down and match the adjoining piece. When you make straight cuts, use a straightedge. Hold the straightedge firmly against the carpeting and use the edge to guide the razor knife. If you are cutting shag carpet, separate the shag piling along the line that you intend to cut. This will expose the backing and make cutting easier.

Seaming involves cutting the carpeting straight and butting the edges together. The edges should fit perfectly, and the pile of the carpeting should blend. First, make the necessary cut. Then match the edges that will adjoin. If the cut is off, you will have to trim the carpeting to fit. Take off small amounts of waste carpeting with the knife; then test the fit.

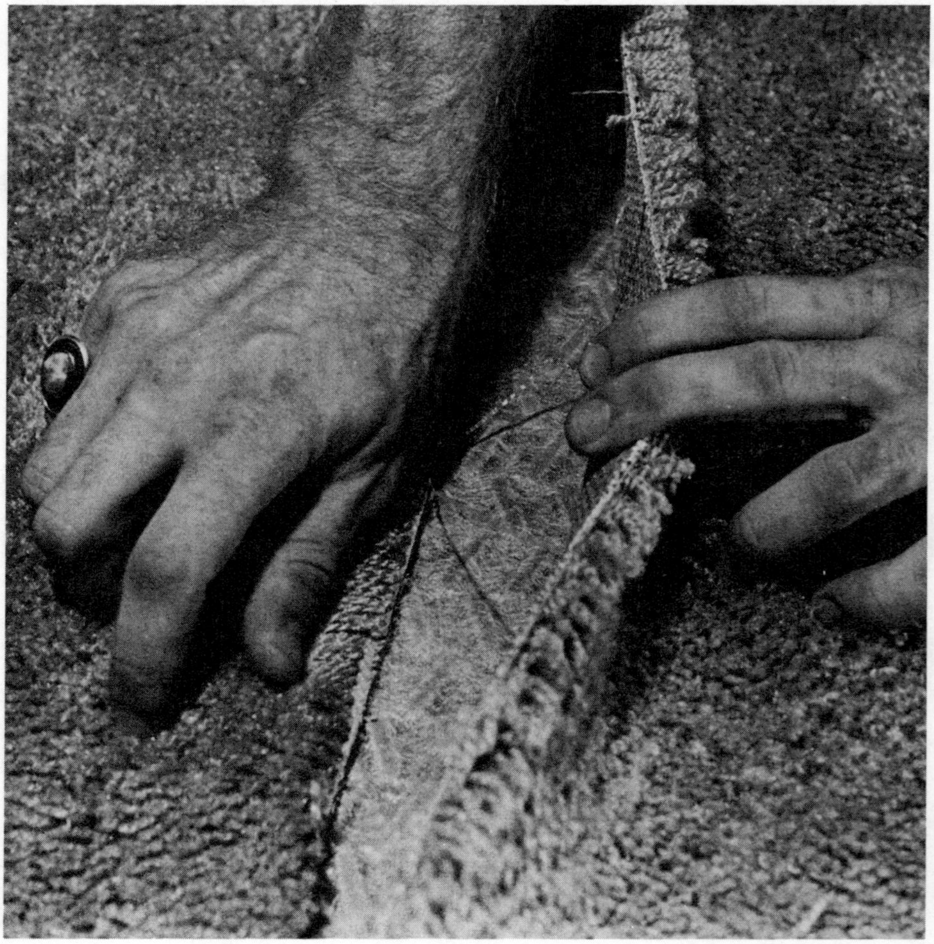

Seaming tape is positioned under the edges of the carpeting to be joined, as shown. Position the tape so that the edges of the carpet split the width of the tape. At this point, double-check to make sure that the pattern of the carpet matches the adjoining pieces and that the nap flows across the seam properly. If your carpeting does not have a pattern in it, you should leave about 2 extra inches at the edges.

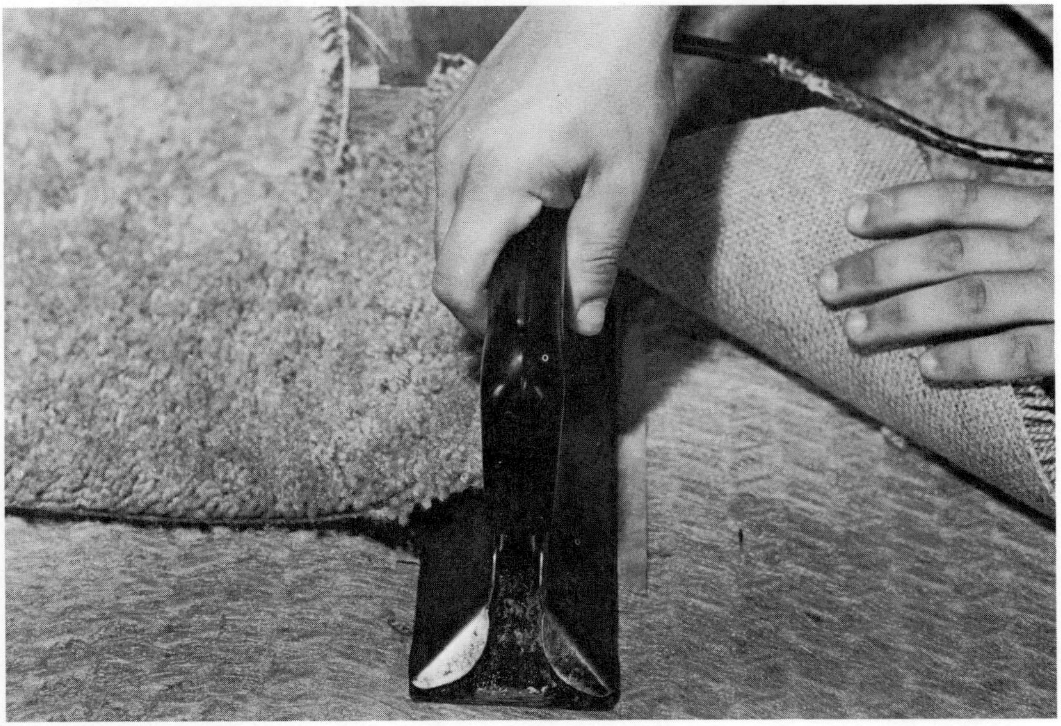

With a hot plate, heat the adhesive on the seaming tape. Then press the carpeting into position. At this point, make sure that the edges match perfectly, the pattern matches, and the nap runs in the same direction. If it doesn't, immediately lift the carpeting off the tape and make corrections. Then start again.

Trim away excess seaming tape. Now take another piece of tape and put it along the edges of the carpeting that butt together in the foreground. When you trim the carpeting to fit this seam, use the edge of the first piece shown to mark the second piece (not shown), since the first piece you have already laid is assumed to be straight.

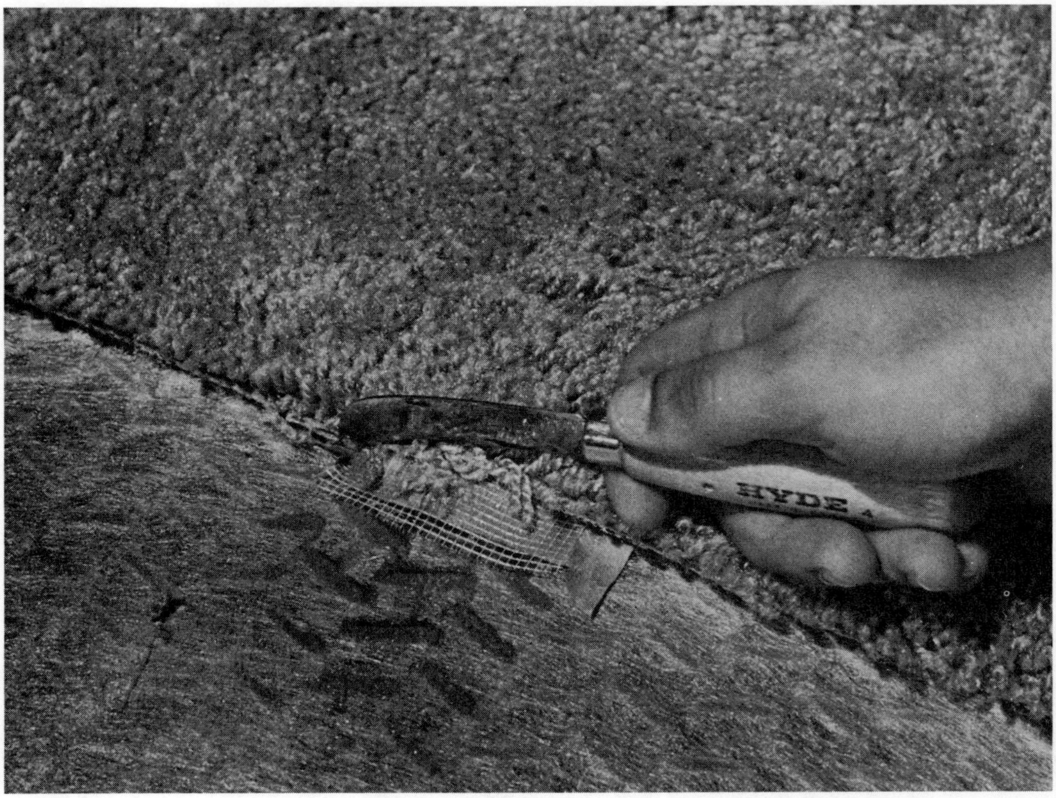

A carpet kit contains tape, seam sealer, solvent, a seam knife and five blades, chalk and chalk line, and a cutting guide. The kit may be used for interior tape-down installation of cushion-backed or indoor/outdoor carpeting. The label on the package explains how to use the kit. The procedures are somewhat the same as for regular seaming tape; you do not need a hot plate.

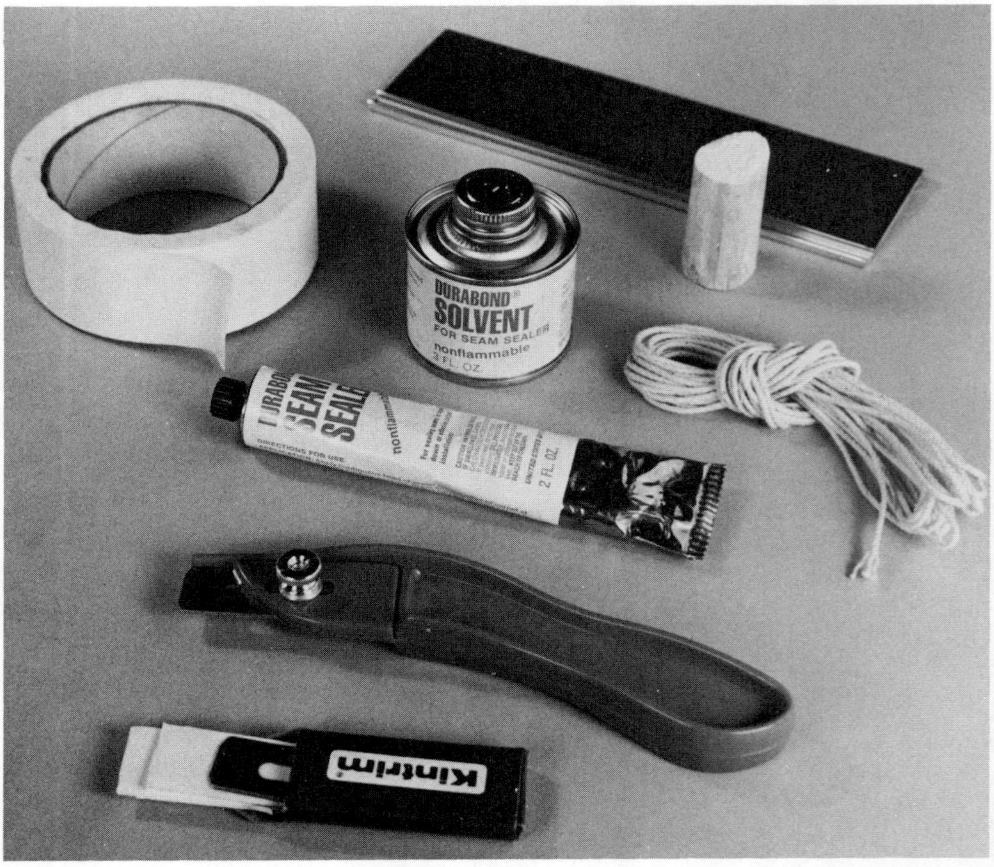

To carpet around corners, notch the carpeting in a V shape, as shown. Note that a 2-inch overlap has been provided for the carpeting. This makes it possible for you to trim the carpet accurately after it has been cut and stretched over the nail strips. If you have to make a double-cut along the edges of the two pieces of carpeting (especially when there is a lot of overlap), lay one piece firmly on the floor. Overlap the other piece. Then cut through both pieces at the same time, using a straightedge to guide the razor knife. If the carpet has a repeated pattern, you will have to overlap the carpet so that the pattern matches.

With a carpet stretcher, stretch the carpet taut against the wall and press it into the nail strips. Do not overstretch the carpet. You don't want it to be too tight; you want it to be comfortably tight and smooth. The stretcher works on a hydraulic action so that you can generate lots of pressure. Simply go around the room stretching and securing the carpeting to the nail strips.

A knee stretcher is used to set the carpeting on the nail strips. To use this tool, trim the carpeting after it has been stretched and attached to the nail strips. Go around the room with a thin trowel or a tool made especially for this (shown), pressing the edges against the baseboard. Tap the padded end of the stretcher with a knee to take up any slack in the carpeting.

Go over the entire floor. If you see any seams that are bunched or do not fit exactly right, lift the misfits and trim them. You probably will have to cut out a portion of the seaming tape and put down a patch.

How to install carpet tile

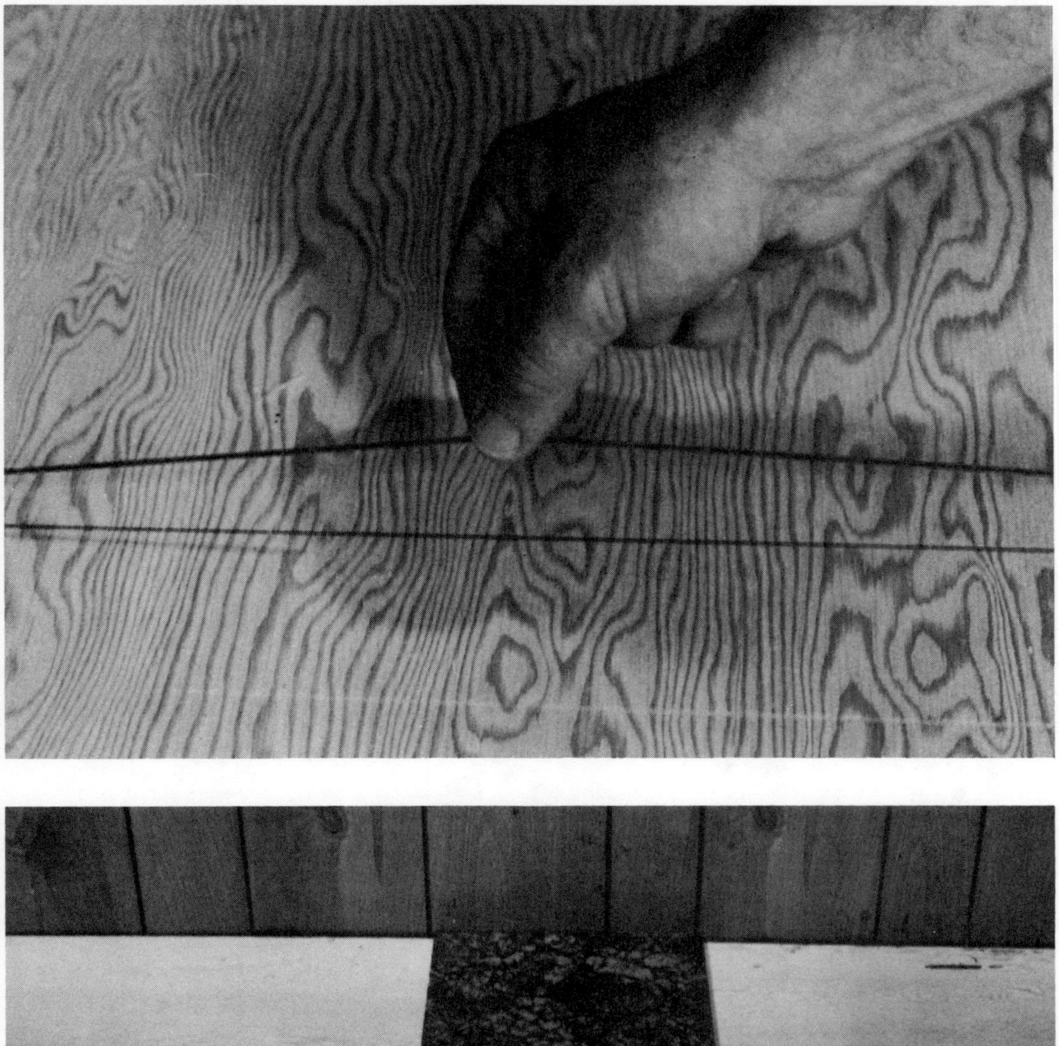

Carpet tile is installed in almost the same way as resilient tile or ceramic tile. We suggest that you read the chapters on resilient and ceramic tile before you start to lay carpet tile. The first step is to prepare the subfloor. It should be clean, smooth, and properly nailed. This procedure is explained earlier in this chapter. Find the center of the room from opposite walls and snap a chalk line in both directions. This will divide the room into quarters. Check the point where the lines cross with a carpenter's square. The lines must form a right angle.

Loosely lay a row of carpet tiles along the lines you snapped. If the distance between the last whole tile and the wall is less than half a tile width, move the center line of the room 6 inches in either direction. This trick will save lots of cutting and fitting later. Work from the center toward the walls.

Arrows on the back of the tiles show the direction the carpet nap runs. The arrows should all point in the same direction as you position the tile on the floor. Do not separate carpet tiles from their backing until you are ready to press them to the subflooring. If the tiles are exposed to the air for long periods of time, the adhesive backing will gather dust and dry out.

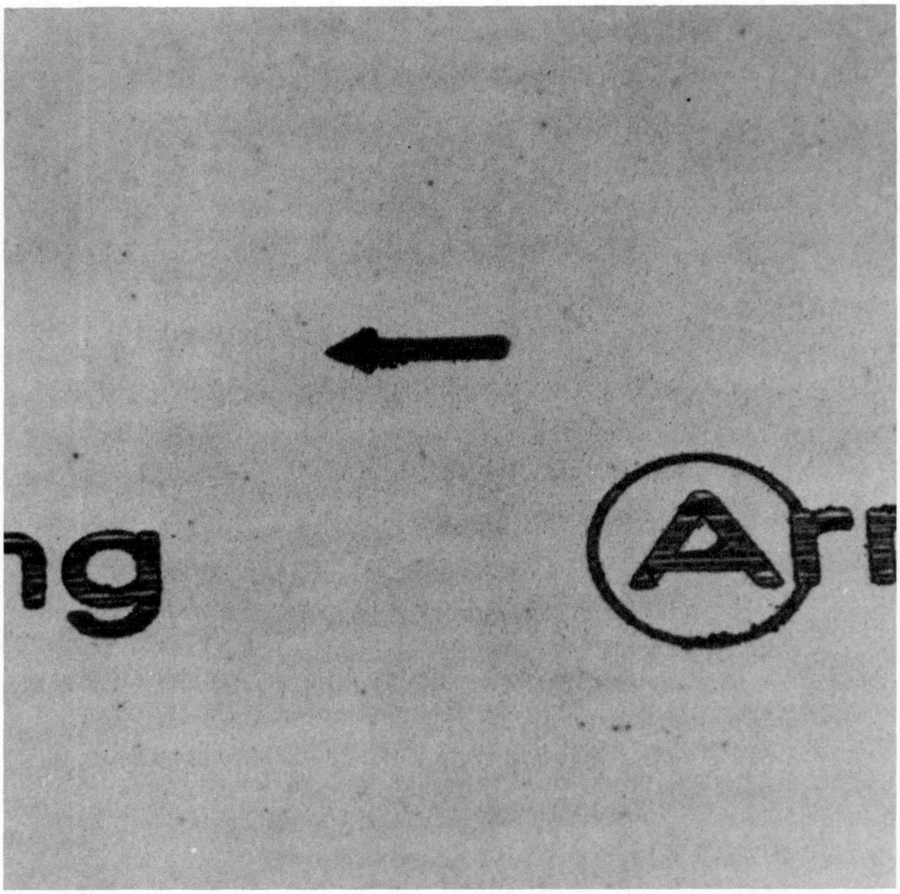

To fit less than full tiles next to the walls, lay a full tile on top of the last full tile you laid. Make sure the edges match perfectly.

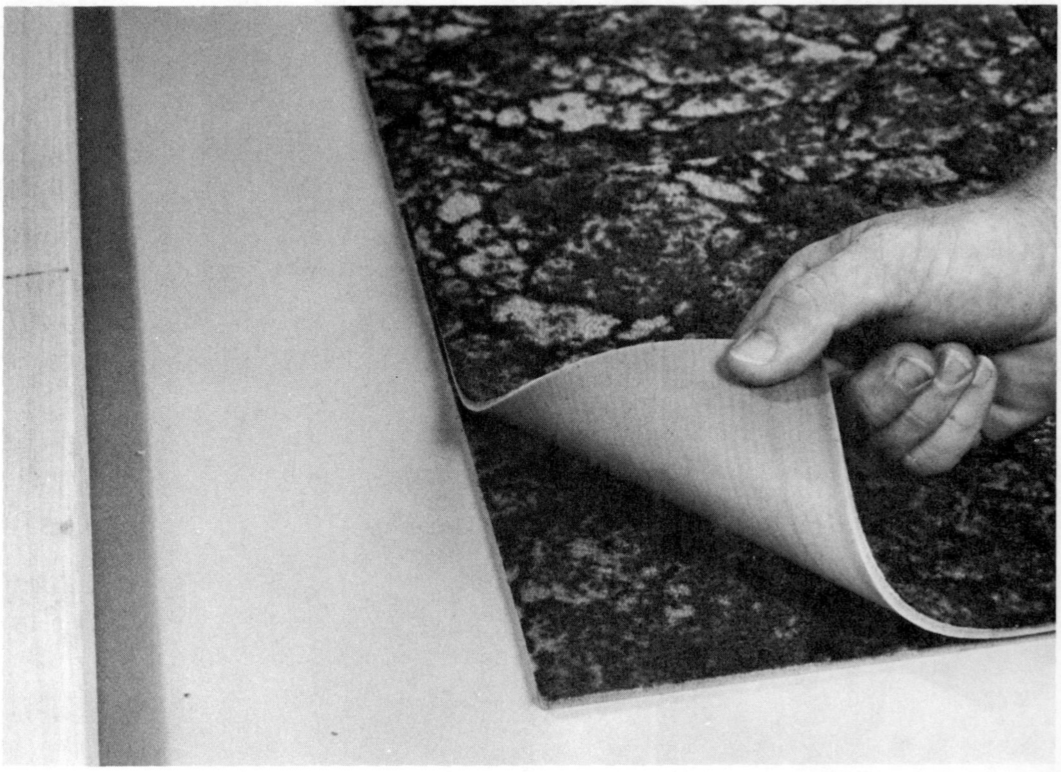

Place another full tile over the second tile, face down, and butt one edge of this top tile against the wall. The top tile should also match perfectly with the edges of the tile underneath it. Then scribe a line for the cut-off point, as shown. Cut the tile to fit with a razor knife or heavy shears. Put the cut edge against the wall when you press the tile into position.

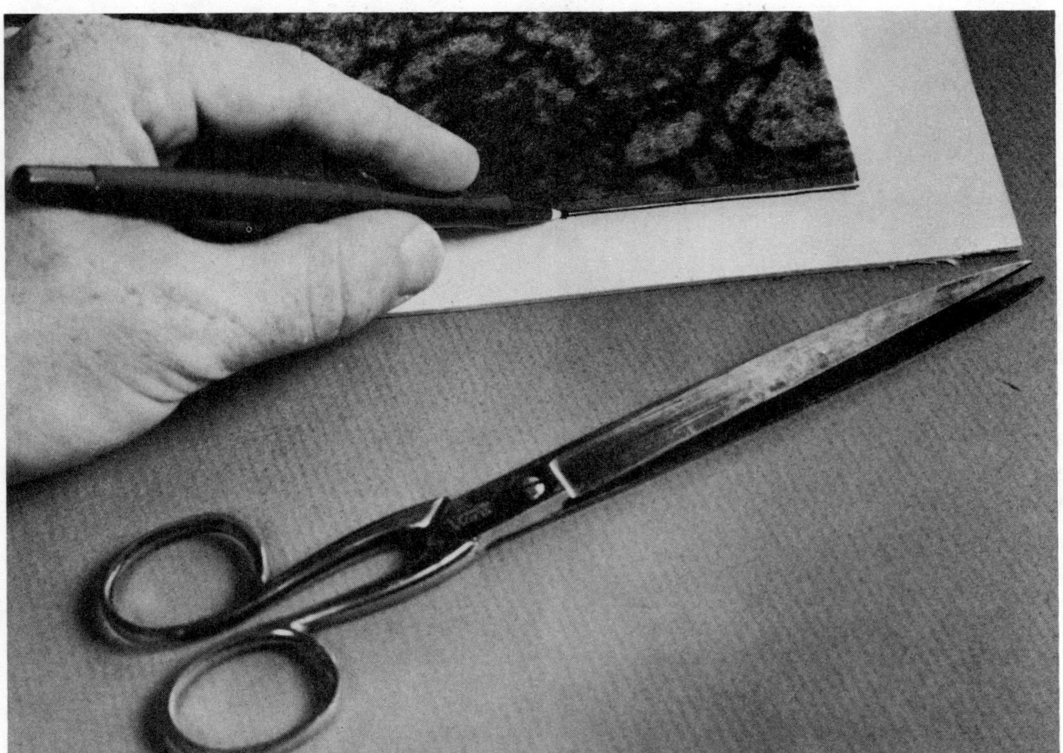

To tile around obstructions, make a cardboard template the same size as the tile. Then fit, cut, and trim the template to match the obstruction. Be as accurate as you can. See the explanation in the chapter "Ceramic Tile and Slate."

Mark the template pattern on the back of the tile, as shown. Then make the necessary cuts with a razor knife or shears. The tile will now fit around the obstruction. Simply press the tile into position. If the tile will not fit, you will have to trim the tile carefully so that it does fit.

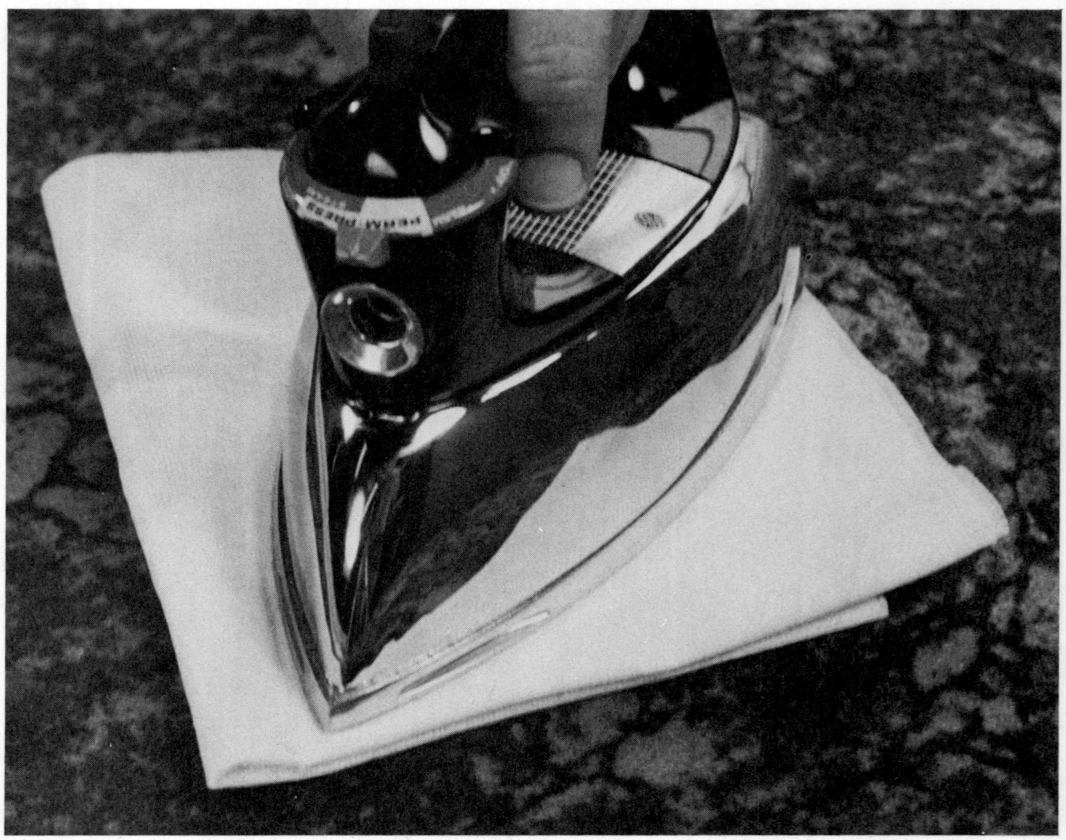

To remove a damaged carpet tile, wet a clean cloth. Then place the cloth over the damaged square. Press the wet cloth with a warm steam iron. Press the iron into the cloth. The steam will loosen the tile so that you can pry it out for replacement. Never put the iron directly on the tile; if you do, the tile may damage the bottom of the iron.

The living area of this home has been opened up to a walled courtyard garden. Shag carpeting used as an area rug brightens both the living room and the former porch that serves as a garden room. Carpeting used in this way should be bound on all sides for a more finished appearance. A sturdy pad under the carpeting will keep it in place. The carpeting contrasts with the hardwood floor and patio blocks to create a pleasing effect.

Floor Repair

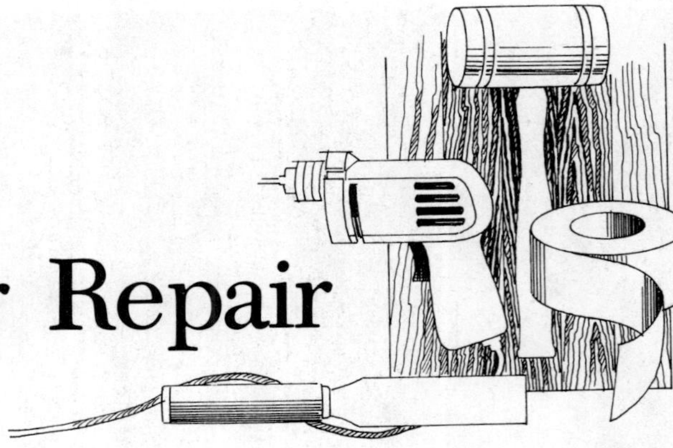

The existing floors and stairsteps in your home may squeak or sag. They may also become damaged by weight, hard knocks, washing, polishing, and water.

These problems may exist even after you install a new floorcovering because they are the result of these conditions:

- Improperly nailed subflooring
- Settling of the house on its foundation
- Damage to the floorcovering surface

Improperly nailed subflooring usually occurs in homes that have carpeting over the subfloor. The carpenters have not used enough nails or fasteners to properly secure the subflooring to the joists. If this is a problem in your home, you will have to remove the carpeting—and probably the carpet padding—and nail the subfloor properly. If the floor squeaks in only a few places, you will not have to go through this procedure. Simply drive a nail through the carpeting and padding into the subfloor and joists. The nap of the carpeting will hide the nailheads.

A house will settle on its foundation normally. You will be able to tell the difference between normal and abnormal settling. If you find large cracks around the baseboards in various rooms, large open cracks around doorways and windows, and large open cracks where two walls join or where the ceiling joins the walls, consult a building contractor. There may be serious damage to the structure of the house.

Damage to the floorcovering surface comes from water, scrapes, and burns. The damaged area may have to be replaced; it certainly has to be repaired.

Squeaks, sags, and other problems on the first floor of the house usually are easy to solve since you can gain access to the joists supporting the floor from the crawl space or basement. Problems with floors on the second and third floors are difficult to solve since the joists are covered with lath and plaster or gypsum wallboard from underneath. Stairsteps are easy to fix from the topside; if you have access to the steps from underneath—such as in a basement—the job is even easier.

If you have a damaged tile floor, you may be able to repair it by removing the damaged tile and replacing the tiles with new ones. The same procedure may be applied to strip flooring, carpet tiles, some hardwood strip flooring, and parquet and block floors.

Damage to carpeting—if it is severe—is difficult to repair and usually requires a professional to make the repairs. It may be best to replace the entire carpeting—especially if the damaged carpeting is old. Reweaving damaged carpeting to renew it may be more costly than replacing the entire carpeting. We recommend that you consult a professional before you replace damaged carpeting.

How to stop floors and steps from squeaking

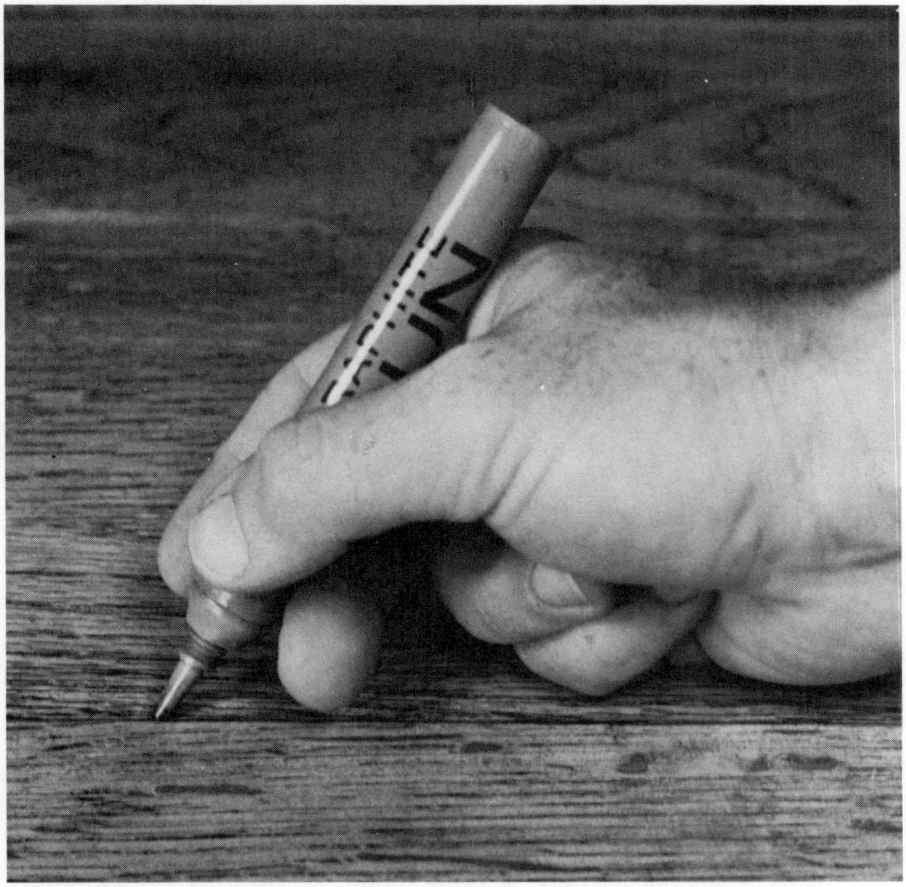

For hardwood strip, parquet, and block wooden floors that squeak, first try putting graphite powder between the joints at the point where the floor squeaks. You can also use liquid graphite oil to stop the squeak. Wooden floors usually squeak because the floor is sagging slightly. This sagging is normal; it is caused by the house settling on its foundation. It may happen in a new home as well as in an older home.

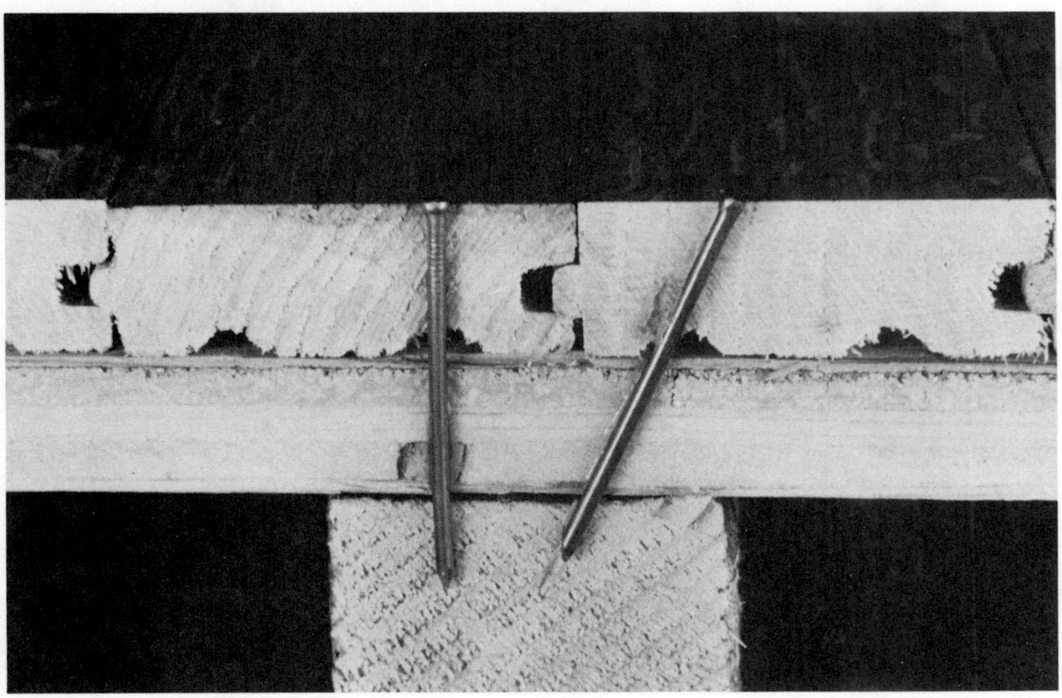

If the graphite treatment doesn't work, try nailing the floor at the point where it squeaks. The best technique is to nail the flooring at an angle (right) through the flooring, underlayment, and subflooring into a joist. If you can't do this, face-nail the flooring (left) into the underlayment and subfloor. Countersink the nailheads with a nail set, and fill the holes with matching wood plastic.

If you have basement access to the floor above, drive a cedar shingle shim between the joists and the subflooring at the point where the floor squeaks. You will need a helper to bounce up and down on the floor above so that you can locate the squeak below. Drive the shim with the side of a hammer. You may have to use several shims at different joist locations to stop the squeak. If the crossbridging between the joists is loose, try nailing the bridging to the joist to stop the squeak.

If the nail or wedge treatment doesn't stop the squeak, try driving a screw through the subfloor at the point where the floor squeaks. You'll need a helper to bounce on the floor above.

Sagging floors often are the cause of floor squeaks. They can also make doors and windows stick and squeak. You can level a sagging floor by inserting a metal jack post under the floor at the sagging point. A jack post must be supported at its base with a wooden timber or a 12-inch-thick concrete pad. The top plate is positioned against the joists where the sag occurs. A 2 by 4 or a 2 by 6 block wedged between the plate and the joists helps distribute pressure from the jack post. Turn the jackscrew tightly against the joists. In a day or two, give the screw a quarter turn. Repeat this every few days until the sag is level. Then leave the post in position or install in its place a Lally column.

Silence squeaky stairsteps by driving a 6d finishing nail at an angle through the back of the tread into the riser. You may have to space a series of nails about 3 inches apart across the back of the tread to stop the squeak. If you can get to the back of the steps, nail through the riser into the back of the tread. Treads and risers are often fabricated from oak. If so, drill pilot holes for the nails, since oak is so hard that the wood will bend the nails. If the steps are carpeted, you do not have to remove the carpeting to drive the nails. The pile in the carpeting will hide the nailheads.

How to replace damaged individual floor tiles

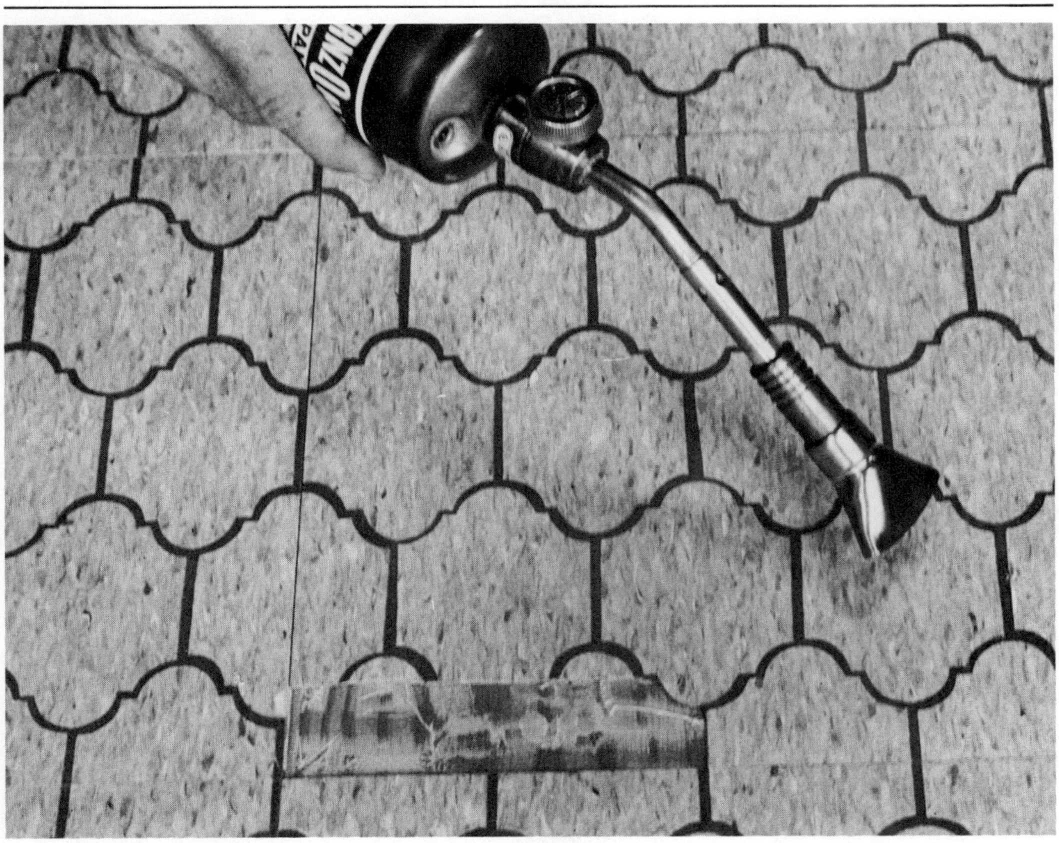

A broken floor tile is easy to replace when you use this method. Heat the tile with a propane torch or an iron, This softens the mastic under the tile. If you use an iron, cover the tile with aluminum foil; the foil protects the bottom of the iron. Test the tile occasionally with a putty knife. When the tile is soft, it is ready to be removed.

With an old chisel or a putty knife, pry out the broken tile. Go easy and nibble out the pieces. This way, you won't damage the underlayment, subfloor, or the tiles surrounding the damaged tile. If you use a chisel, work with the beveled edge down—not the flat side. You don't need a hammer to hit the chisel; hand pressure is enough.

Remove old, dried mastic from the underlayment or subfloor. A chisel is the best tool for this job. Then apply mastic to the underlayment or subfloor with a notched trowel, as shown. Use plenty of mastic and spread it evenly in the void. Make sure there is adequate adhesive next to the adjoining tiles.

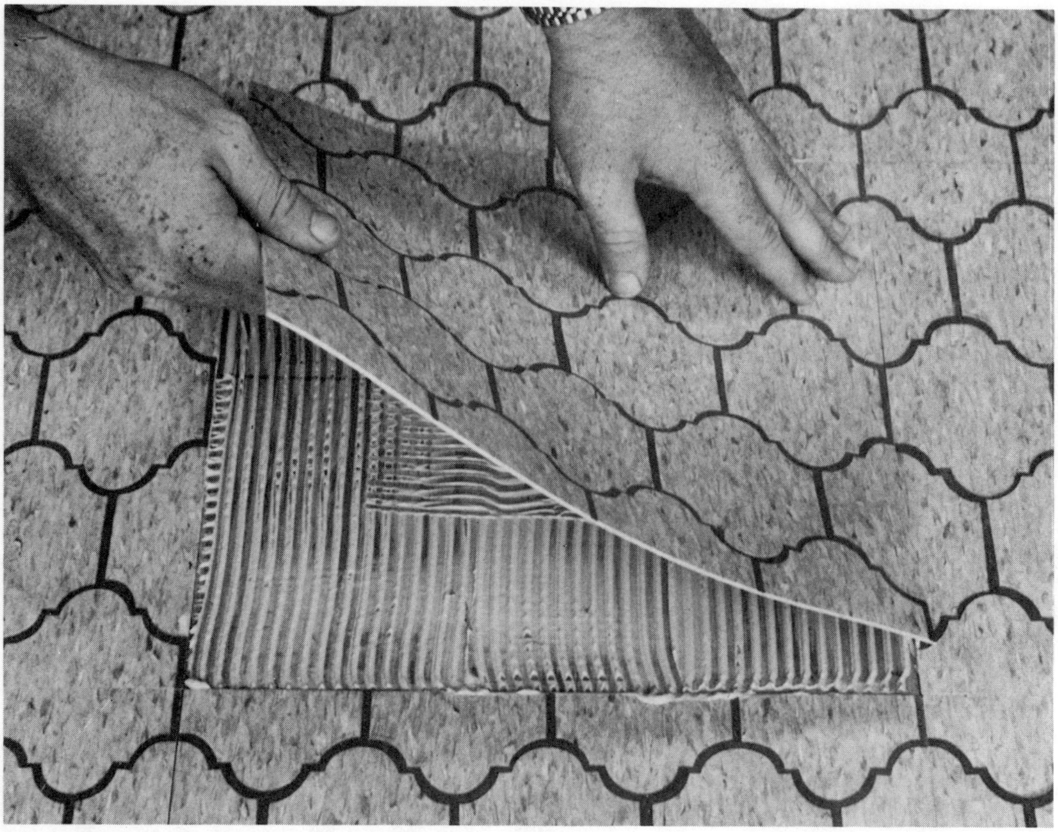

Carefully press the replacement tile into the adhesive. Butt the edges of the replacement tile tightly against two of the adjoining tiles; then press the entire tile into position. The adhesive probably will ooze from under the replacement tile and onto the surrounding tiles. This is normal. But immediately wipe up any excess mastic with a soft cloth. Take a rolling pin and roll the replacement tile into the adhesive. Or cover the tile with a piece of hardboard or plywood and step on it. Then remove the board and wipe up any excess adhesive.

Cover the replacement tile with hardboard or plywood, as shown, and weight the board. Let the job sit for a day or so until the adhesive has an opportunity to harden. Then remove the weight and board. The replacement tile will probably appear cleaner than the surrounding tiles. If so, give the replacement tile several coats of floor wax. The wax buildup helps blend the new tile with the old surrounding tiles.

When there are small dents in the floor tile, you may not have to replace an entire tile. Clean the damaged tile with a knife. Then shave small particles off a matching tile with a razor knife. Make the particles as tiny as you can. Now mix the particles with a quick-drying clear varnish. Try to make a paste of the particles. With a putty knife, trowel the paste into the dents. Smooth the surface. When the paste has dried, buff the patch smooth with fine steel wool that has been soaked in linseed oil. Let this dry. Then wax the patched spots and the rest of the floor.

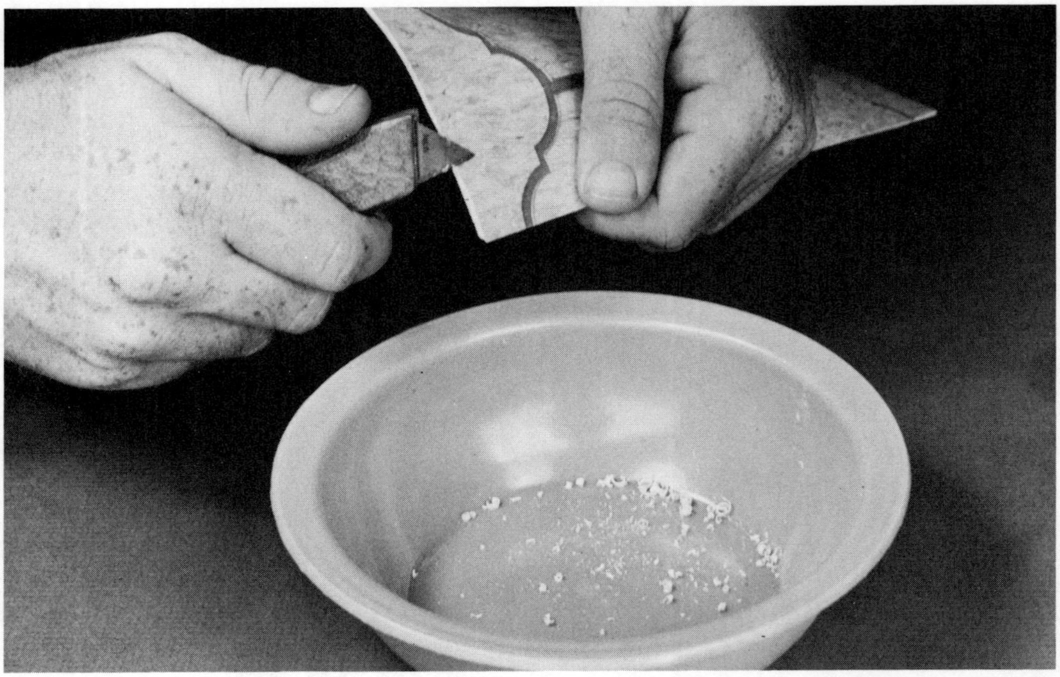

How to patch damaged sheet flooring

The patch for sheet flooring must match the pattern of the flooring. Therefore, the patch must be cut large enough to cover the damage and to match the surrounding flooring. The size of the patch is determined by the pattern. You will have to experiment with the patch before you do any cutting.

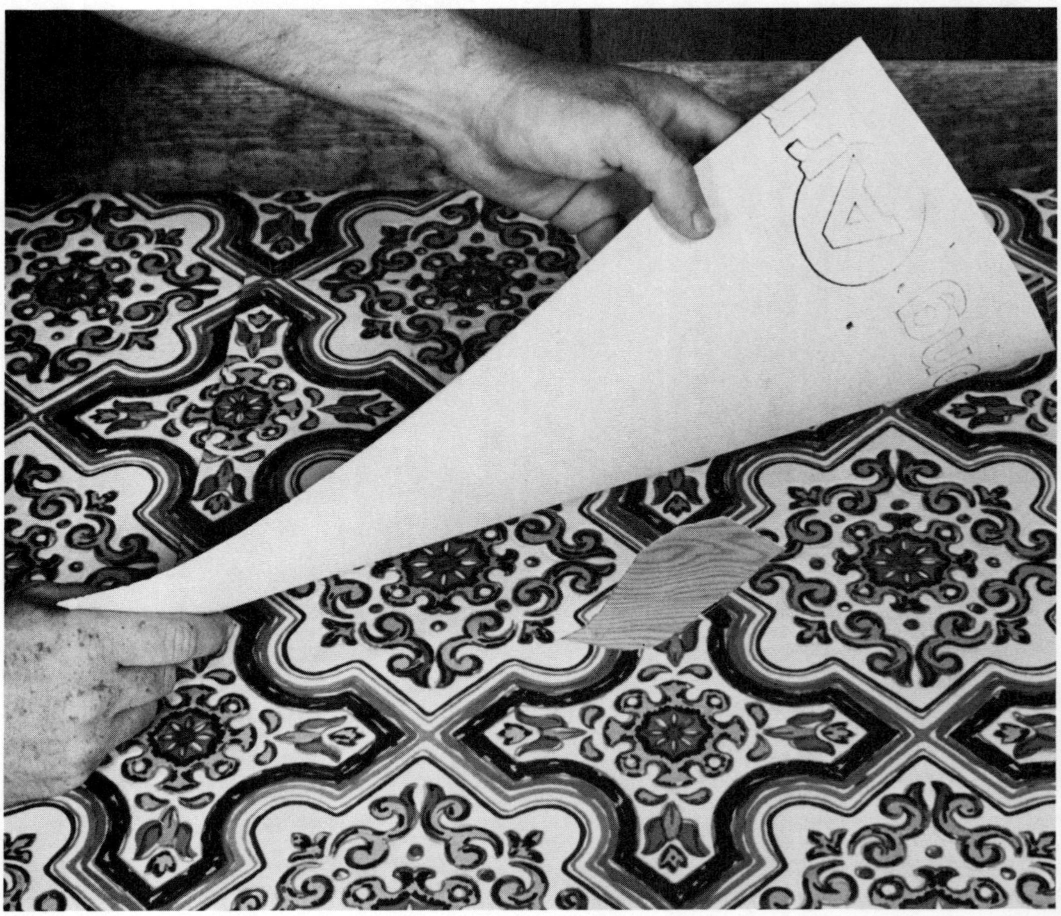

Cut the patch to match the pattern. Then carefully match the patch to the floor and anchor the patch into position with masking tape. Use enough tape so that the patch will not slip when you start to cut.

With a straightedge, cut through the patch and the flooring around the damaged area. A short length of aluminum angle makes an excellent straightedge. Or use a carpenter's framing square. Try to cut the patch as uniformly as possible; a fairly large patch is easier to handle than a small one.

After you make the cuts, remove the patch and the damaged portion of the flooring underneath the patch. Clean the underlayment or subfloor. Then fit the patch in the void. You may have to trim the patch slightly so that it fits and matches the pattern perfectly. Daub adhesive on the back of the patch, or spread the adhesive on the underlayment with a notched trowel. You need only a thin coating of adhesive.

Press the patch into place after the adhesive is applied to the patch or the underlayment. Then roll the patch with a wallpaper seam roller or a rolling pin. This will press the patch firmly into place. Weight the patch by covering it with a piece of hardboard or plywood. Let the patch dry for a day before you remove the weight.

How to replace damaged strip flooring

Drill a series of holes across the width of the damaged strip or strip flooring. You will have to replace this strip from cross joint to cross joint. Put several series of holes down the length of the strip. The number of holes depends on the length of the strip. Drill just to the underlayment; don't force the drill into the underlayment or subflooring.

With a sharp chisel, split the flooring between the series of holes. The more you split the flooring, the easier it will be to remove the flooring. As you tap the chisel into the strip, remove any excess wood. You won't be able to remove the entire damaged strip, but removing portions of it makes it easier to split the remaining piece of flooring. Use the same technique for removing and replacing damaged porch flooring.

Remove the damaged strip. You may have to pull several flooring nails so that the void left by the damaged strip is clean and smooth. Measure and mark the replacement strip that will fill the void. Make sure the new strip butts tightly against the strip it is to adjoin. If the damaged strip butts against a wall, you will have to remove the base trim to take off the damaged strip.

Remove the lower half of the groove side of the new strip. Use a sharp chisel for this and work carefully so that the edge will be perfectly square. The trick is to shave off as much of the groove as possible with a chisel. Plane the edge smooth and square with a jack plane. Make sure the blades of both tools are sharp; oak is a hard wood and dulls cutting edges fast.

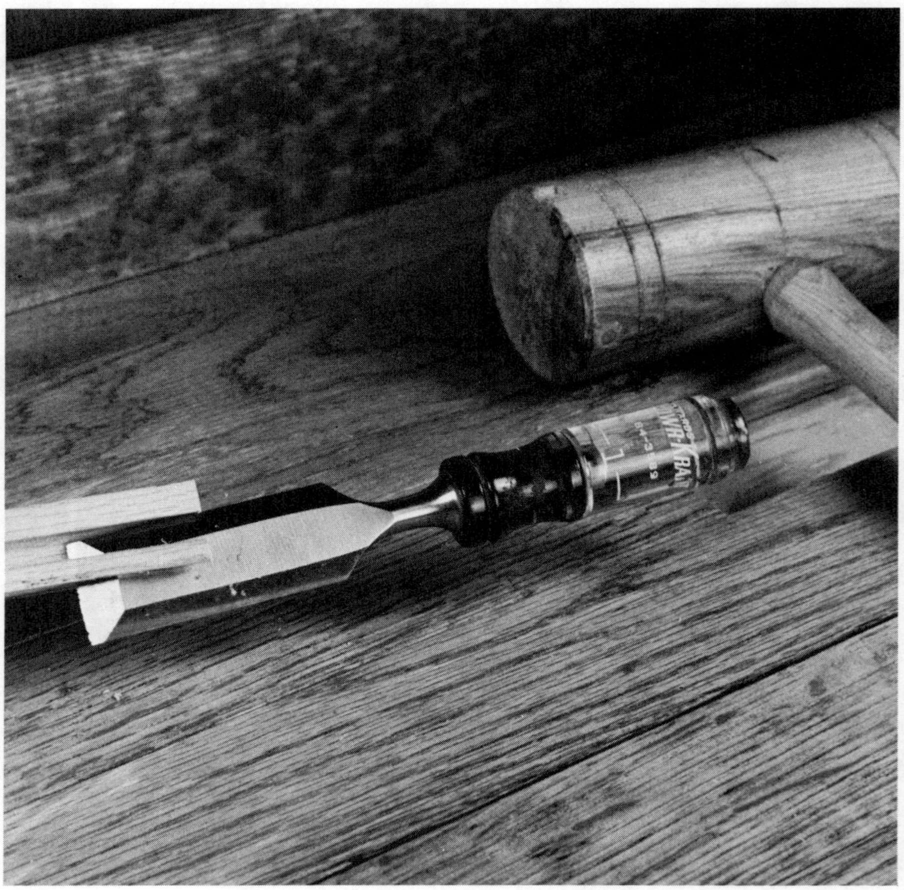

Tap the replacement strip into position. The tongue of the replacement strip fits into the groove of the adjoining strip. Do not force the replacement strip into position. If the strip tends to bind, remove it and cut away the excess wood with a jack plane. Forcing the replacement strip into the void may split the wood. When you're finished with this, the replacement strip should fit tightly in the void, and the joints should be tight.

Drill pilot holes for nails in the face of the replacement strip. You need pilot holes because the flooring is hard and difficult to nail. Drill through the strip but not into the underlayment and/or subfloor. We recommend that you use 6d finishing nails for the flooring. Space the nails about 12 to 15 inches along the length of the replacement strip.

Nail in the replacement strip. Do not make hammer tracks on the flooring. If you are replacing a strip of porch flooring, nail the replacement strip to the joist. Porches are usually not installed over underlayment and/or sheathing. Mark the joist positions on the replacement strip so that you can drive the nails accurately.

Countersink the nailheads with a nail set. Then fill the holes with wood plastic or wood putty. The plastic or putty may be stained to match the flooring. When the filler is dry, lightly sand the top surface of the filler and the replacement strip. Then shellac, stain, varnish, or wax the replacement strip to match the surrounding floor.

How to repair scratches in hardwood strip flooring

To repair small dents in hardwood strip flooring, you do not have to replace the entire damaged strip. Clean the dent with a chisel. Then fill the dent with wood plastic or plastic wood. When the filler dries, sand the patch smooth and finish it with shellac, stain, varnish, or wax. You can buy wood plastic in many different wood tones. This material is available at home center stores and building material and hardware outlets.

Shallow scratches in hardwood strip flooring—and parquet and block flooring—can be erased without replacing the damaged flooring. Use fine steel wool and a liquid solvent cleaner. Rub the steel wool in the direction of the wood grain. Do this until the scratch disappears. Then clean the area with a good detergent, rinse, and wax the floor.

If the scratch is deep, you may be able to use wood plastic or wood putty to fill it. First, sand the scratch with a medium-grit abrasive. This helps to hold the filler in the scratch. With a putty knife, trowel in the filler so that it protrudes slightly above the surface of the flooring. The filler tends to shrink slightly when it dries. Let the filler set for several hours. Then lightly sand the spot smooth so that it is flush with the surrounding floor.

Refinish the patched area. Use orange or clear shellac thinned with alcohol. Make a mixture with 3 parts shellac to 1 part alcohol. Or, if the floor is varnished, use varnish thinned slightly with turpentine. After the finish has dried, it may be glossy. If so, you can dull the finish to match the rest of the floor with fine steel wool.

Index

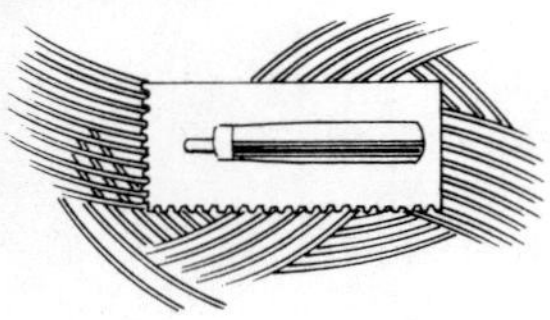